The Economics of Land Degradation

DEVELOPMENT ECONOMICS AND POLICY

Series edited by Franz Heidhues, Joachim von Braun and Manfred Zeller

Vol. 66

PETER LANG

Frankfurt am Main · Berlin · Bern · Bruxelles · New York · Oxford · Wien

The Economics of Land Degradation

Toward an Integrated Global Assessment

Ephraim Nkonya / Nicolas Gerber /
Philipp Baumgartner / Joachim von Braun /
Alessandro De Pinto / Valerie Graw /
Edward Kato / Julia Kloos / Teresa Walter

PETER LANG
Internationaler Verlag der Wissenschaften

Bibliographic Information published by the Deutsche Nationalbibliothek
The Deutsche Nationalbibliothek lists this publication in the Deutsche Nationalbibliografie; detailed bibliographic data is available in the internet at http://dnb.d-nb.de.

ISSN 0948-1338
ISBN 978-3-631-63082-2

www.peterlang.de

Foreword

Land and soils have been undervalued since long at a global scale and in many world regions. Their true values beyond market prices, including ecosystem services, and the costs of land degradation being a loss of natural resources, are not appropriately accounted for. This undervaluation is coming to an end quickly but not smoothly in recent years. As food and agricultural commodities become more scarce and their prices show an upward trend, the price of land is also increasing and the value of soil fertility as a foundation for future production is now more appreciated. That price trend will not change for a while as climate change will add to the pressure on fertile land, and growing demand for biomass due to population growth, income growth and the expanding search for bio-based rawmaterials.

Due to these trends there is growing interest in land investments by the private and public sectors. Still, policy action is lacking and a policy framework for action is missing. To trigger action, one step needs to be taken beforehand: we need to raise awareness about what is at stake. In pursuit of that an assessment of the state of affairs regarding land degradation and the costs of related inaction is called for. This book aims at providing a basis for that, with the hope to facilitate the needed action.

It is pointed out in this volume, that globally, land degradation affects 1.5 billion people who depend on degrading areas, and it is closely associated with poverty, with 42 percent of the poor living in degraded areas. The indirect effects of degradation have global impacts beyond degraded areas through food price increases and potentially also migration and conflicts.

The book points at deficiencies in knowledge about land degradation, its measurement, and the needs for clarification and harmonization of the concepts and data sources. An assessment of the costs of land degradation must consider the formal and informal rules governing the land use decision.The conceptual framework proposed to compare the costs of action against land degradation versus the costs of inaction may help to address these deficiencies. Underlying causes of land degradation are the reasons why unsustainable land management practices occur, including ecology, land shortage, weak institutions, lack of know how, and economic pressures.

All the important actors of land degradation, such as land users, landowners, governmental authorities, and industries need to be involved in strategies for reduced land degradation. These actors, besides the research communities of a broad range of disciplines, are the target groups of this volume.

Joachim von Braun Franz Heidhues Manfred Zeller

Acknowledgments

The authors are grateful to Rattan Lal, professor of soil science at the School of Environment and Natural Resources, and director of the Carbon Management and Sequestration Center, both at The Ohio State University; to Franz Heidhues, professor emeritus of rural development policy and theory at the University of Hohenheim; and to Paul Vlek, director at the Zentrum für Entwicklungsforschung (ZEF; Center for Development Research) in the department of ecology and natural resources management at the University of Bonn for their timely and constructive review of an earlier draft of this report. We are also grateful for the insightful comments from the staff members of the Convention Project to Combat Desertification (CCD Project), Deutsche Gesellschaft für Internationale Zusammenarbeit (GIZ) GmbH; staff members of the United Nations Convention to Combat Desertification; Richard Thomas of the United Nations University Institute for Water, Environment, and Health; and the participants at the partnership meetings held in Bonn on December 15, 2010.

The financial support of this joint project of International Food Policy Research Institute (IFPRI) and ZEF by GIZ on behalf of the German Federal Ministry for Economic Cooperation and Development is gratefully acknowledged.

The authors take responsibility for all content, errors, and omissions in the report.

Contents

List of Tables

List of Figures

Executive Summary

Introduction

Since the publication of the Report of the Brundtland Commission (*Our Common Future)* in 1987, and the consequent Earth Summit on sustainable development, global attention on natural resource scarcity and degradation has been increasing, because of climate change and rising food and energy prices. This awareness, in turn, has led to growing interest in land investments by the private and public sectors. Despite this interest, however, land degradation has not been comprehensively addressed at the global level or in developing countries. A suitable economic framework that could guide investments and institutional action is lacking. This study aims to overcome this deficiency and to provide a framework for a global assessment based on a consideration of the costs of action versus inaction. Thus, a type of Stern Review (Stern 2006) for land degradation (LD) is aimed for on the basis of this study. The urgency of land degradation problems, the increased value of land, and new science insights all suggest that the time is ripe for a global assessment of the economics of land degradation (ELD).

Land degradation is taking place in all agroecological zones. It has long affected the world and it is increasingly considered a global problem, as its extent and impacts is increasingly affecting environmental and social vulnerability. Land degradation, described by the Global Land Degradation Information System (GLADIS) as the "reduction in the capacity of the land to provide ecosystem goods and services over a period of time" (Nachtergaele et al., 2010, 9) is occurring worldwide and depends on the linkages between several natural and socioeconomic determinants. Humid areas seem to have a higher share of the global land degradation than initially thought (Bai et al. 2008a). According to the Global Land Degradation Assessment (GLADA), land degradation is increasing, with almost one-quarter of the global land area being degraded between 1981 and 2003. The most severely affected areas are Africa south of the Equator, Indochina, Myanmar, and Indonesia. Desertification describes land degradation in drylands,[1] which cover about 41 percent of Earth's land surface and which are home to more than 38 percent of the global population (Reynolds et al. 2007). In this report, which aims to set the platform for a global and comprehensive assessment of LD, the more inclusive term of *land degradation* is preferred, as it implicitly includes desertification. Measured as net primary production (NPP), without taking atmospheric fertilization into account, land degradation caused a total loss of 9.56 x 10^8 tons of carbon

---

1 Drylands include arid, semiarid, and subhumid areas with an aridity index (AI), measured by the ratio of precipitation to potential evapotranspiration, between 0.05 (hyperarid) and 0.65 (dry subhumid).

between 1981 and 2003, which amounts to $48 billion in terms of lost carbon fixation.[2] Globally, LD affects 1.5 billion people who depend on degrading areas, and it is closely associated with poverty, with 42 percent of the very poor living in degraded areas, compared with 32 percent of the moderately poor and 15 percent of the nonpoor (Nachtergaele et al. 2010).[3] Each of the concepts and data sources in recent global approaches (GLADA and GLADIS) has its relevance, but all have serious deficiencies, as they do not explicitly relate to economic productivity and human well-being (for example, NPP and Normalized Difference Vegetation Index [NDVI]).

The extent of degraded and degrading areas affects large numbers of people and leads to significant social and economic costs, thus raising the questions: In which way is it worth taking action against LD? Where, when (earlier rather than later?), and at what costs? To make informed decisions about how to prioritize action, it is essential to know the social and economic costs linked to the current and future status of LD. In addition, it is necessary to know what kind of actions can be taken against LD and their specific costs. In that sense, this report aims to deliver an extensive review of the current literature on the quantification and mapping of LD, its effects and driving forces, the economic valuation of its different costs, and the way it is embedded in the institutional and policy context.

Based on this review, a general conceptual framework is proposed to compare the costs of action against LD versus the costs of inaction. The framework presented here and partly illustrated with case studies will prepare the groundwork for a global assessment on the costs of LD.

Although recognizing the differences among desertification, and land degradation, this report conceptualizes the economic valuation of LD in terms of its association with the change in productivity and with changes in the provision of ecosystem services and the human benefits derived from them. Ecosystem services comprise provisioning (agricultural output, fuelwood, fresh water, and so on), supporting (such as soil formation, nutrient cycling), regulating (such as water and climate regulation), and cultural services.

A Global Economic Assessment of Land Degradation

For the assessment of land degradation, it is important to identify its driving forces. Proximate causes of land degradation can be split into biophysical causes—such as topography, climate conditions and change, and natural hazards—and unsustainable land management practices. Underlying causes of land degradation are those that indirectly affect the proximate causes. —In

2 Using a shadow price of carbon of $50 per ton (British treasury, February 2008)

3 Classification of poverty into nonpoor, moderately poor, and very poor was based on infant mortality rates (see Nachtergaele et al. 2010).

general, underlying causes can be described as the reasons why unsustainable land management practices occur. Underlying causes include land shortage, poverty, migration, and economic pressures and their drivers. These different causes often contribute simultaneously to the severity of land degradation. The resulting level of land degradation then determines the specific effects of land degradation in terms of the loss in ecosystem services and the benefits that humans derive from them. This loss gives rise to on- and off-site costs and to direct and indirect costs.

On-site and *off-site costs* are related to the location where these costs arise (at the plot, within the watershed, or even globally) and whether they are considered in land use decisions. *Direct* and *indirect costs* relate to the consequences of land degradation. Indirect costs include impacts on land markets and other agricultural input markets that have further effects on food prices or on other sectors of the economy. Through multiplier effects, the whole economy is affected. Indirect costs are thus also related to poverty, food security, and other global socioeconomic issues.

The economic valuation of the various terrestrial ecosystem services needs to be rooted in the framework of total economic value. This ensures that all values (use, nonuse, and option values) are considered. A comprehensive assessment needs to capture all changes in ecosystem services attributed to LD. The potential problem of double-counting needs close attention, as ecosystem services are not easily split up into particular benefits that can be valued and then aggregated (Balmford et al. 2008).

Reviewing the literature on valuation studies of LD reveals that the main focus has been on impacts on agricultural productivity, especially its direct costs. Yet this is only part of the provisioning services of terrestrial ecosystems, and indirect costs, such as the human suffering caused by LD and the loss of biodiversity, are still poorly researched. Most of the studies on the costs of land degradation (mainly limited to soil erosion) give cost estimates of less than 1 percent up to about 10 percent of the agricultural gross domestic product (GDP) for various countries worldwide. Because agricultural productivity is simultaneously determined by a variety of ecosystem services, the productivity change approach is an important tool in evaluating the costs of LD. The decrease in agricultural productivity represents an on-site cost. In addition, off-site costs need to be appropriately accounted for, because they are high. For example, globally, the cost of the siltation of dams is about $18.5 billion. Special existing valuation techniques can be applied to that end. Furthermore, the indirect costs on the economy (national income), as well as their socioeconomic consequences (particularly poverty impacts), must be accounted for, at least qualitatively. These costs are difficult to measure and require empirically sound models that are able to integrate the relevant linkages in the economic system. Not all is understood yet in terms of the linkages between LD and its effects on human welfare and across the whole economy. Only then can a comprehensive cost assessment of global LD be achieved. This assessment has to be conducted at the margin, which means that costs of small changes in the level of LD, which may accumulate over time, have to be identified. Bringing together the different cost and value types to fully assess total costs and benefits over time and their interactions can be done within the framework of cost–benefit analysis and mathematical modeling.

The assessment of the costs of LD must identify and consider institutional and political arrangements, because they represent the formal and informal rules governing the economic production and thus influence land use decision. It is also essential for the analysis to identify all the important actors of LD, such as land users, landowners, governmental authorities, and industries, as well as identify how institutions and policies influence those actors. Transaction costs and collective versus market and state actions are to be considered. The institutional economics is particularly important in the assessment of LD when it comes to the definition and design of appropriate actions against LD, as well as of the inaction scenarios serving as a benchmark.

Implementing the Cost of the Action versus Inaction Approach

Appropriate actions against LD consist of institutional and policy measures and sustainable land management techniques. The latter range from traditional conservation measures, such as the application of fertilizer and manure, to installments of hedgerows or terraces and agroforestry, to the breeding and cultivation of crops with higher nutrient-use efficiency and drought-tolerance. Appropriate mixes of institutional, policy, and land use measures can be designed into different action scenarios; a cost assessment is necessary for each of these scenarios. The benchmark for comparing the costs of action is the business-as-usual cost of inaction scenario, which estimates the impacts of maintaining the current level of action taken against LD.

This report's proposed framework for a global cost assessment of action and inaction against LD must incorporate some challenging issues, some of them warranting further research:

- *LD is highly site specific.* Because a variety of site-specific exogenous and endogenous factors influence the processes and levels of LD, cost assessments need to be based on the microlevel, where LD can actually be observed. Existing global measurements in the form of NPP, NDVI, and so on cannot capture the actions that have already taken place to mitigate LD. Hence, though they capture interesting information, their use to assess costs of mitigation is limited.
- *Site-specific analysis must be upscaled.* The microlevel information on LD causes, effects, and costs must be translated into global-level information in order to estimate the global costs. However, clear

difficulties exist in applying the methods discussed in the empirical research (case studies) for measuring the direct costs of LD in a global assessment. More research is required to refine scenario definitions, upscaling from site-specific case studies to the national and international level and determining how to capture comprehensively direct and indirect costs of LD.

- *LD is time dependent.* LD takes place and has impacts through time—most often in a nonlinear, dynamic way. How LD and its impacts evolve over time and how the corresponding costs change can be difficult to forecast and assess. The choice of the discount rate, reflecting the time preference and the time horizon for a global assessment, is an important point to consider.
- *Rural poverty is a cause and outcome of land degradation.* Typically, in poor rural areas, people have a high time preference, which means they attach more value to the present and discount the future at a higher rate than do most other people. They are also strongly risk averse. Because land conservation measures usually require high initial investments, with payoffs arising much later, this attitude can lead to more land degradation. As many of the poor are highly dependent on the increasingly degrading land with decreasing productivity, they may slide ever deeper into poverty, which leads to even more underinvestment in conservation measures and to more land degradation (though alternative pathways with successful collective action have also been observed). Therefore, the cost estimation must include the poverty effects of LD and their potential for correcting processes. The dynamics of such self-correcting processes against LD are not well understood and pose a challenge to the valuation of the costs of action.
- *There are multiple linkages between LD and climate change.* The land surface is part of the climate system in many ways, and climate change, in turn, influences land degradation processes and drought. Many aspects of the LD–climate change relation are still unknown. The potential for self-accelerating dynamics to lead to tipping points of disasters exists among these relations. However, specific actions against LD hold the potential for win–win–win situations—that is, measures taken to decrease land degradation that simultaneously improve the productivity of the land and decrease the climate change impacts of land uses. More research based on sound modeling is required.

Despite the numerous challenges, a global assessment of the costs of action and inaction against LD is possible, urgent, and necessary. This assessment can and should be combined with remote sensing and geographic information system (GIS) analysis of the appropriate data to link those data to existing global land degradation monitoring tools and evidence-based and evidence-checked modeling. Mapping as an outcome of the cost of action versus inaction approach allows for the identification of focus areas for actions against LD. These areas could be areas in which the costs of action are low or the costs of inaction are high. They could also be areas in which the impact on humans is high, with severe poverty and food security consequences. In that respect, maps are great tools to assist in the analysis of the data and to inform actors involved in the decisionmaking processes at all levels of the fight against LD. Such mapping, however, needs to be done both ex post and ex ante with the appropriate modeling tools.

This study provides a framework for a global assessment; part of that framework includes systematically selected country studies. In a comprehensive assessment, the country studies must cover all settings representatively. In this study, only indicative case studies are provided. The country case studies suggest that the cost of action is lower than the cost of inaction for seven of the eight cases, even though the costs of degradation are only defined in terms of decreased crop yields. These results suggest the need to explore other reasons for not taking action—for example, lack of access to markets and rural services, such as agricultural extension services.

To conduct a global economic assessment of LD, major aspects of the complex biophysical and socioeconomic factors must be taken into account. A great deal can be learned from past global-level studies, and this new effort to strengthen the economics of LD should definitely build on strengths of past studies and the growing interdisciplinary collaboration. Because past studies have tended to be specialized, they have produced rigorously analyzed results that have informed researchers of other disciplines. Although this work can be built upon, a redesigned process for shaping an integrated assessment is needed. In that respect, lessons can be learned from other global environmental assessments, such as the Economics of Ecosystems and Biodiversity (TEEB) or the Intergovernmental Panel on Climate Change's (IPCC's) work and communication on climate issues. An organizational design for the global assessment is proposed here; this proposed design is simple and would operate at low transaction costs. We propose an institutional setup for the implementation of a global ELD assessment in which all stakeholders can meet and interact for the benefits of global action and investment against land degradation. It would also include potential science partners involved in relevant activities, without confessing their distinct rates as decisionmakers and knowledge providers. An open-consultation process across all different groups of the institutional setup would be a worthy initial phase to the global assessment, as well as a continuation of the dialogue process of which this study is a part. Given the variety of people and interests involved, it is crucial that the global assessment not only be based on country representation, but also on a people-representation principle. Minorities in a given area (for example, pastoralists and nomadic groups) might represent major stakeholders in the LD debate. This inclusion would not only give an important role to nongovernmental organizations (NGOs) and civil societies in the global assessment process, as they are the voice of specific minorities, it would also ground-truth the links between LD biophysical effects and community and human effects.

Data availability is a major challenge of studying ELD, and partnership networks in data collection and research will help address this challenge. Remote sensing, GIS, and modeling based on the cost of action versus cost of inaction framework are required. Access to data from past case studies will also help attain data that require experiments or detailed information. Assessment of the causes and impacts of LD will also require collecting data that cannot be captured by satellite imagery, as well as the collection of on-site data for ground-truthing satellite data. Collecting such data at the regional or global scale is expensive and will require partnering with existing national bureaus of statistics, which regularly collect household surveys. Finally, the global assessment of LD and its costs should systematically review the institutions and stakeholders involved in LD programs. A new framework for action is needed to facilitate a comprehensive global assessment of LD. A cost of action versus inaction framework would facilitate in guiding and accelerating the process.

To capture the main findings of the study the reader can look at the summary sections of this report's core sections (2, 3, 4, and 5), Section 7, and the conclusions (Section 8).

1. Introduction

1.1 Background and Choice of Approach

Land degradation (LD) has affected the world for centuries, and it is reported to be increasing in many parts of the world, with negative consequences on the productivity of the land and its ability to provide ecosystem services. However, greater attention has turned to these problems only over the past two to three decades. The establishment of the United Nations Convention to Combat Desertification (UNCCD) in 1994 showed that these problems presently receive worldwide recognition. It also became obvious that further studies to assess LD are necessary in order to understand their causes, scale, and diverse effects.

The terminology *LD* used in this report captures comprehensively the terms *land* and *soil degradation* as well as *desertification*. Having reviewed a number of studies, we found that *land degradation* and *desertification* were often used interchangeably. Strictly speaking, desertification refers to land degradation in the arid and semiarid zones; however, the term conveys the idea of land being converted into deserts, which is not true for all land degradation processes. Further, as a considerable share of land degradation takes place in humid areas, a global assessment must take into account land degradation in all climate zones. Hence, we chose to use the more comprehensive and correct term of land degradation in this report.

In order to provide policymakers with evidence-based recommendations on how to deal with LD in the future, we must include the quantification of LD and its consequent social and economic costs. It can be assumed that LD is costly; however, not enough research has been done to assess its true costs or the benefits of preventing LD or of rehabilitating degraded lands.

This report responds to the request of the German Federal Ministry for Economic Cooperation and Development (BMZ) and the UNCCD for a pilot report that scopes the science about LD economics. This report prepares the groundwork for a more comprehensive initiative on the economics of LD and its integrated global assessment.

Research into the economics of LD (ELD) studies advocacy, mobilization of finances, and appropriately targeted investment to prevent or mitigate LD. One example is climate change. The *Stern Review on the Economics of Climate Change* (Stern 2006) made a strong impact, especially its conclusion that investing (now) in the mitigation of greenhouse gas emissions makes more economic sense than facing the (future) costs of failing to do so. A similar review on the economics of biodiversity loss was the Potsdam Initiative, which was initiated by the G8 in 2007. That initiative resulted in The Economics of Ecosystems and Biodiversity (TEEB) study (Balmford et al. 2008). The present report shares similarities with the TEEB study's objectives, broad methodology, and specificities of topics covered. First, this study is intended as a preparatory work for future, more comprehensive work on the science of LD and its drivers, costs, and mitigation. Second, LD is of similar importance to humans in that it affects the provision of terrestrial ecosystem services and therefore the benefits these services provide that affect human well-being.[4] Thus, the economic value of these ecosystem services is a necessary input into the evaluation of the costs of LD and its mitigation. Unfortunately, and despite considerable advances in the economic valuation of ecosystem services in recent years, there are still many gaps in such valuations, including in the economic valuation of specific ecosystem services and in the geographic coverage of such valuation. Drylands—areas where LD can be particularly detrimental to the well-being of their inhabitants—are one type of area in which this coverage has been quite poor.[5]

In the present report, the conceptually covered ecosystems (and the services and benefits they provide to humans) include all terrestrial ecosystems, including anthropogenic ecosystems—that is, ecosystems that are heavily influenced by people (Ellis and Ramankutty 2008). These anthropogenic ecosystems include agroecosystems, planted forests, rangelands, urbanization, and so on. Meanwhile, a majority of the literature investigating the impacts of LD, its costs, and mitigation does so within the realm of agroecosystems. This fact is, of course, reflected in the outputs of this report. Nonetheless, agroecosystems (defined as spatially and functionally coherent units of agricultural activity) are strongly linked to conventional ecosystems, or those ecosystems with minimum or no human influence. The link between anthropogenic and conventional ecosystems is through the provision of certain ecosystem functions, such as support and regulation services (for example, nutrient cycling, climate regulation, and water purification). Hence, to also assess these impacts, LD and its costs should be assessed across all terrestrial ecosystems.

Changes in terrestrial ecosystems are used to define land degradation or land improvement. Thus, we start by discussing the terrestrial ecosystems and how they are treated in this study.

---

4 The report does not assess the intangible values attached to terrestrial ecosystems. This does not mean that the authors deny the intrinsic value of terrestrial ecosystems as a justification for conserving them. As Balmford et al. (2008) pointed out, "An evaluation of the economic value of [terrestrial ecosystems] is not incompatible with this conviction. Indeed, if the results of such evaluation are that conservation results in a net economic gain, then that simply adds an economic argument against [the degradation of terrestrial ecosystems], alongside the moral argument. If the results are that conservation of [terrestrial ecosystems] incurs a net economic loss, then that will provide the net size of the bill for conserving [terrestrial ecosystems]" (7).

5 This is, for instance, reflected in the relatively small number of case studies reviewed under TEEB that are linked to drylands.

1.2 Terrestrial Ecosystem Services

Recognizing that the term *land* refers not only to soil, the UNCCD (1996) defines land as "the terrestrial bioproductive system that comprises soil, vegetation, other biota, and the ecological and hydrological processes that operate within the system" (Part 1, Article 1e). Land as an ecosystem provides ecosystem services that have been defined by Costanza et al. (1997) and "represent the benefits human populations derive, directly or indirectly, from ecosystem functions" (253).

The Millennium Ecosystem Assessment (MA 2005) links ecosystem services to human welfare and concludes that degradation of ecosystems reduces human welfare. Ecosystem services are categorized into supporting, provisioning, regulating, and cultural services (see Box 1.1).

Box 1.1—Ecosystem services

- **Supporting services:** Services that maintain the conditions of life on Earth—soil development (conservation/formation), primary production, nutrient cycling
- **Regulating services:** Benefits obtained from the regulation of ecosystem processes—water regulation, pollination/seeds, climate regulation (local and global)
- **Provisioning services:** Goods provided—food, fiber, forage, fuelwood, biochemicals, fresh water
- **Cultural services:** Nonmaterial benefits obtained from the ecosystem—recreation, landscapes, heritage, aesthetic

Source: MA 2005b.

The concept developed by MA (2005b) was instrumental in illustrating the importance of ecosystem services to human well-being. LD can be assessed as a loss in ecosystem services provided and in the resulting lost benefits to humans. The concept of ecosystem services, however, is not perfectly suitable for framing the economic valuation of land resources, as pointed out by various studies (Boyd and Banzhaf 2007; Wallace 2007; Fisher, Turner, and Morling 2009). Therefore, we will also draw from the concept of on- and off-site effects, which has been widely used in literature related to land degradation and which provides a coherent framework for economic valuation of the costs arising from the different effects.

A comprehensive coverage of the issue of LD must address its impacts on the entire range of ecosystem services and their benefits to human well-being. Starting with the various services offered by agroecosystems is only one step in that direction, as

- other ecosystems offer benefits to humans that clearly cannot be neglected (see TEEB reports for illustrations); and
- dynamic interactions between the costs of LD and the opportunity costs of changes in land use—that is, changing the nature of ecosystems and, therefore, of the services and benefits provided for humans—should be considered.

As stated in the objective that follows, this study will review the current knowledge and the state of the art of analytical approaches to LD; will propose a pragmatic approach for assessing LD; and will describe the nature of the partnership required to implement a global ELD study.

1.3 Objectives

The objectives of this report are as follows:

1. To assess the current knowledge and the state of the art of analytical approaches to LD and to identify knowledge gaps; the discussion focuses on
    - the causes and driving forces of LD;
    - the effects of LD on land productivity, including changes in the provision of terrestrial ecosystem services;
    - the social and economic costs of LD; and
    - the costs and benefits of enhancing land productivity and (re-) establishing ecosystem services.
2. To propose a methodology for an integrated evaluation of the aforementioned costs based on LD assessments at global, regional, and local levels.
3. To describe the nature of the partnerships required for regional and global assessments and to recommend potential partners.

The outputs are as follows:

1. A wide range of state-of-the-art knowledge in the area of the economic costs of land degradation and sustainable land use, building, in particular, on the existing literature on the valuation of LD and its effects vis-à-vis the valuation of sustainable land management. The primary interest is in the comparison of cost of action versus cost of inaction.
2. A review of the available methodologies for assessing the social and economic costs of LD, with recommendations for selecting a methodology for choosing representative case studies whose results could be extrapolated to comparable sites in order to obtain an estimate of the global cost of LD and the global benefits of preventing or mitigating LD.

3. Requirements and advice on the methodology for carrying out representative case study assessments and the global study, as well as a consideration of the nature and type of partners and partnerships needed.

1.4 Conceptual Framework

This report first seeks to assess the existing knowledge and the state of the art of analytical approaches to LD, as well as the costs of LD, and to identify knowledge gaps in this regard. Second, the report proposes a methodology for prioritizing across possible geographic areas of intervention. The approach we recommend is based on an assessment of the costs of investing in the effective prevention and mitigation of LD against the costs of the loss in terrestrial ecosystem services and the benefits those services would deliver to human beings—that is, the cost of action versus inaction. Finally, this report reviews and advises on the methods that attempt to identify activities and areas in which investments would have the highest expected returns.

Framework: Confronting Action versus Inaction

The conceptual framework used for the analysis in this report is presented in Figure 1.1. The different elements of the conceptual framework are then discussed in more detail in later sections. Scenarios A and B represent two states of the world—one in which action is taken to prevent or mitigate LD or restore land after LD took place (Scenario A), and another in which no action against LD has been taken (Scenario B). The round-edged gray boxes represent physical elements of the system under consideration that have direct relationships, and the black arrows denote the direction in which the impacts flow. Proximate causes of land degradation are those that have a direct effect on the terrestrial ecosystem. The proximate causes are further divided into biophysical proximate causes and unsustainable land management practices.

Examples of the biophysical proximate causes of land degradation are topography, which determines soil erosion hazard, and climatic conditions, such as rainfall, wind, and temperature, all of which could prevent or enhance land degradation. For example, rainstorms could trigger flooding and soil erosion. Rainfall could also enhance land cover and therefore prevent soil erosion. Unsustainable land management, such as deforestation, soil nutrient mining, and cultivation on steep slopes, directly leads to land degradation. The underlying causes of land degradation are those that indirectly affect the proximate causes of land degradation.

Figure 1.1—Conceptual framework for assessing the costs of action versus the costs of inaction regarding LD (with net present value outcomes).

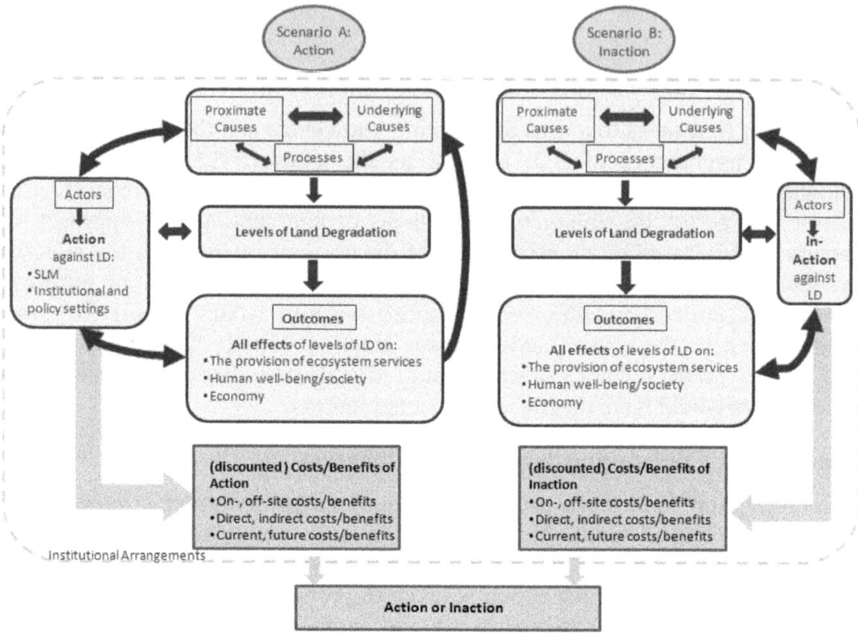

Source: Author's creation.

Note: SLM = Sustainable land management.

For example, poverty could lead to the failure of land users to invest in sustainable land management practices. Similarly, policies that enhance investment in land management, such as payment for ecosystem services in China, which enhanced tree planting on steep slopes in northwestern China in the western provinces, can affect the proximate causes of land degradation (erosion on slopes in this case). Population density could lead to intensification (Boserup 1965; Tiffen, Mortimore, and Gichuki 1994) or to land degradation (see, for example, Grepperud 1996), depending on other conditioning factors. The causes of land degradation are discussed in more detail in Section 2. The level of land degradation determines its outcomes or effects—whether on-site or off-site—on the provision of ecosystem services and the benefits humans derive from those services. If actions to halt or mitigate LD are taken, the actors involved in halting mitigation are determined by the causes of land degradation that need to be addressed, by the level of land degradation, and by its effects. Actors can then take action to control the causes of LD, its level, or its effects. Many of the services provided by ecosystems are not traded in markets, so the different actors do not pay for negative or positive effects on those ecosystems. The concept of externalities refers to the costs and benefits arising from the production or consumption of goods and services for which no appropriate compensation is paid (for example, off-site effects such as sedimentation and indirect effects such as migration and food insecurity). The value of such externalities is not considered in the farmer's land use decision, which leads to an undervaluation of land and its provision of ecosystem services and, in suboptimal levels, of LD.

The green square boxes deal with the economic analysis that is carried out, and the light green arrows show the flow of information that is necessary to perform the different elements of the global economic analysis. A decrease in the provision of terrestrial ecosystem services and their benefits has direct economic costs to humans, such as decreased food security and increased food prices (via the role of markets). These costs are quantified in the two parallel green boxes. In addition to affecting the stream of benefits derived from ecosystem services, LD also has indirect effects of importance to humans. For instance, it affects land and other agricultural input markets, thus affecting their prices and the prices of the goods produced. Further, the impacts on the agricultural market (on any sector that depends directly on terrestrial ecosystem services and benefits) have intersectoral, economywide effects that are passed to other sectors by what economists call *multiplier effects*. Thus, these direct and indirect effects of LD can widely affect poverty and national income and thus have far-reaching socioeconomic consequences. The indirect effects and their costs are also analyzed in the two parallel green boxes. Ideally, all indirect and off-site effects should be accounted for in the economic analysis to ensure that the assessment is from society's point of view and includes all existing externalities, in addition to the private costs that are usually considered when individuals decide on land use. How the different concepts described so far—externalities, private and social costs, and on-site, off-site, direct, indirect, current, and future costs—relate to each other is depicted in Figure 1.2.

Similarly, actions against LD have direct benefits and costs—the costs of specific measures and economywide indirect effects—that is, opportunity costs. In other words, resources devoted for these actions cannot be used elsewhere. Thus, mobilizing those resources to prevent or mitigate LD affects other sectors of the economy as well. The comparison of the values obtained in the two parallel dark gray boxes—the costs of action versus inaction against LD—is the purpose of the global analysis.

Figure 1.2—Costs and benefits—Concepts

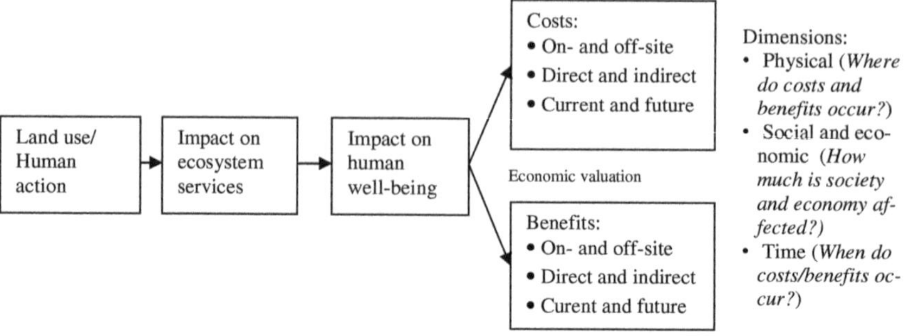

Source: Author's creation.

Institutional arrangements—or the "rules of the game" that determine whether actors choose to act against LD and whether the level or type of action undertaken will effectively reduce or halt LD—are not represented in the conceptual framework. Nonetheless, it is crucial to identify and understand these arrangements in order to devise sustainable and efficient policies to fight LD. For example, if farmers overirrigate, leading to salinization of the land, it must be understood why they do so. As an illustration, it may be that institutional arrangements, also referred to as *distorting incentive structures*, make it economically profitable for farmers to produce as much crops as possible. Missing or very low prices of irrigation water in irrigation schemes act as such an incentive in a misleading institutional setup. The rules of the game play a role in determining the causes, outcomes, and actions to prevent or mitigate LD, but they are too complex to be portrayed accurately in Figure 1.1. Hence, they are captured as a dashed circle enclosing and affecting everything within it. How exactly institutions affect land use decisions is the subject of Section 4.

Market Failures for Most Ecosystem Services

Markets for most of the regulating and supporting services do not exist. The lack of market prices for ecosystem services means that the benefits derived from these goods (often public in nature) are usually neglected or undervalued in decision making. Many ecosystem services are public goods, which means they are usually open in access. Public goods are nonrival in their consumption—that is, consumption of the good by one person does not diminish the amount consumed by another person. The public-good nature of many ecosystem services prevents correct pricing of the resource and creates the need for economic valuation and policymaking (Freeman 2003).

Land use decisions rarely consider public benefits and mostly focus only on localized private costs and benefits. Benefits that occur after a long-term horizon, such as that from climate regulation, are frequently ignored. This neglect leads to a systematic undervaluation of ecosystem services, because values that are not part of financial or economic considerations are somehow ignored. The failure to capture the values causes non optimal rates of land degradation. To adequately account for ecosystem services in decision making, the economic values of those services have to be determined. In Section 4, a number of economic valuation techniques that can be used are discussed. However, attributing economic values to ecosystem services is challenging, due to many unknowns and actual measurement constraints. As economic values linked to the number of (human) beneficiaries and the socioeconomic context, these services depend on local or regional conditions. This dependence contributes to the variability of the values (TEEB 2010). The precondition for an economic valuation of the ecosystem services provided by land is knowledge regarding the state of land resources and their ecosystem services. Although recent global approaches to the assessment of land degradation have greatly contributed to the knowledge of worldwide land degradation and its relevance for human development, they are still subject to criticism regarding the accuracy of the results on the regional and local level. The gaps related to the measurement of LD and, more specifically, to the provision of ecosystem services challenge their economic valuation. As TEEB (2009) indicates, a global framework that identifies a set of key attributes and then monitors these by building on national indicators could help answering this challenge.

Valuation of Marginal Changes in LD

Typically, land degradation is a slow-onset problem. Therefore, the valuation of the different costs must be conducted at the margin (Stern 2006; Balmford et al. 2008). It is the costs induced by marginal (relatively small) changes in the level of LD that must be evaluated through both their impacts on the provision of terrestrial ecosystem services and their benefits. There is a need for marginal analysis because land degradation does not represent binomial events—that is, it does not account for the presence or absence of land degradation, but rather for events taking place slowly in different orders of severity (for example, low or severe degradation). Desertification is not a binomial event either; rather it slowly covers a wide range of land degradation levels in dryland areas. Hence, the costs of actions can be measured against these small changes: How much does it cost to generate a small improvement of LD-affected lands?[6] Further, the definition of marginal changes in LD is usually determined by changes that can be brought about by specific land conservation policies (for example, planted

6 This approach can pose problems when there are abrupt ("nonlinear" in economic jargon) changes in the relationship between the level of LD and the provision of terrestrial ecosystem services. Therefore, particular attention must be paid to the study of drivers of these relationships.

hedgerows to prevent erosion) or sometimes by specific levels of mitigation policies (for example, fertilizers to compensate for low soil nutrients). "[The] valuation of [the] marginal changes is anchored into specific 'states of the world' generated from counterfactual scenarios where a specific policy action is either adopted or not" (Balmford et al. 2008, 7-8). The states of the world in a specific area are determined by economic geography, land use, population distribution, and land conservation and LD mitigation strategies, as well as external factors, such as climate. Actions against LD, based on empirical evidence, are used to generate different states of the world using projection scenarios and simulations.

Site-Specific Valuation

Naturally, the impacts of specific actions on specific types of LD, such as terracing against erosion, are highly site and case specific. Hence, the description of the state of the world is usually a description of the state of the site. Due to the time restrictions under which this report was compiled, it was not possible to be exhaustive in the list and types of sites that were covered. Rather, acknowledging the need to be site specific in the valuation of the costs of LD (action versus inaction), we review and propose methodologies to be subsequently applied to a wide range of case studies. The case studies presented in this report serve the purpose of illustrating these methodologies.

Time Dependence and Discounting

Degradation of an ecosystem may not translate directly or immediately into a loss of services. Ecosystems can take up to a certain level of degradation and then start to decline rapidly (TEEB 2009). The impacts of specific LD processes and of the actions used to mitigate them are felt through time, in a way that is most often nonlinear. For instance, whereas terracing might have a direct and stable affect on erosion levels (when all other components of the state of the site are kept constant), the impact of afforestation on nitrogen cycling is clearly time dependent. Similarly, the impacts of specific land degradation processes on the provision of ecosystem services vary across time. For instance, erosion has a nonlinear impact on crop yields, as the erosion of top soil depletes nutrients mostly early in the process, thus rapidly affecting crop yields; further erosion, however, shows limited impact on soil. Hence, an analysis of erosion might show no impact on agricultural yields, even though or because the crucial resource (top soil) has already been depleted. With such dynamic processes and links, we must ideally value ecosystem services in a nonstatic way, aggregating the economic value of terrestrial ecosystem benefits through time. The impacts of marginal changes in LD are then expressed in terms of their impact on the discounted value of a stream of terrestrial ecosystem benefits. The benefits are discounted in order to compare the value of the aggregation of present and future benefit streams, expressed in terms of their value at a common point in time—the present. The cost of inaction is equal to the difference between the sums of the discounted ecosystem benefits and their direct and indirect effects under the action and inaction scenarios (the former should be higher than the latter, unless LD is broadly beneficial to human well-being and LD-preventing investment is not competitive with alternative investments in human well-being in the long run). The cost of action is equal to the discounted costs of the different LD prevention, mitigation, or restoration costs, including their off-site, indirect, and future costs. The cost of preventing land degradation will be much smaller than the cost of rehabilitating already severely degraded lands (see Figure 1.3). Hence, costs of action will increase the more actions against LD are delayed.

Figure 1.3—Prevention, mitigation, and rehabilitation costs over time

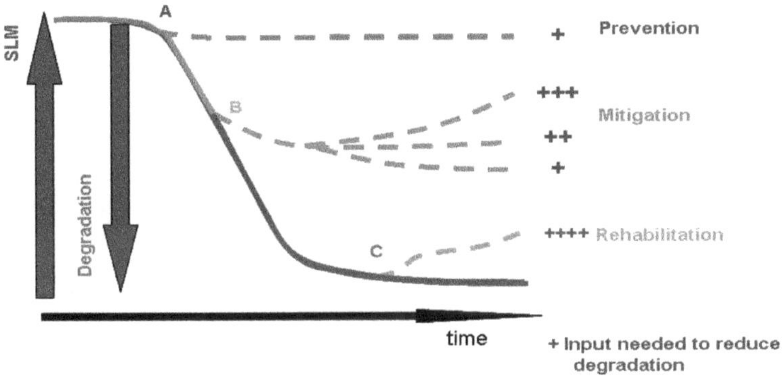

Source: Schwilch et al. 2009.
Note: SLM = sustainable land management

The choice of discount rates and time horizon is crucial, because the size of the discount rate, as well as the length of the considered time horizon, can radically change the results. Discount rates relate to people's time preference, with higher discount rates indicating a strong time preference and attaching a higher value to each unit of the natural resource that is consumed now rather than in the future. Assume a future benefit of sustainable land management of $10,000. If the discount rate were 3 percent, then the present value of the benefit of $10,000 earned in 10 years from now would only be $7.441. A zero discount rate refers to no time preference at all and the equal valuation of a unit of the resource consumed in the present or in the future. High discount rates, therefore, tend to discourage investments that generate long-term benefits and favor those that create short-term benefits but significant long-term costs. The social discount rate, from the perspective of society as a whole, is usually much lower than the discount rate of individuals; from a societal point of view, the value of a unit of the natural resource consumed by this generation or the next is the same, everything else being kept constant, whereas this does not hold from an individual's perspective. Further, discount rates can actually differ across individuals, based on socioeconomic factors, attitude toward risk, and uncertainty. When applying the methodology presented in this report, the issue of discounting and time and planning horizons needs to be clearly elaborated. What discount rate to set is primarily an ethical question. Following the argument that present generations have the moral obligation to protect the interests of future generations would lead to rather low discount rates. Actions of today's generation impose intergenerational externalities, such as loss of biodiversity, on future generations. On the other hand, today's investments are likely to increase the wealth of future generations, which allows them to more easily address environmental protection, at least as long as environmental damage is reversible. The existence of inflation, time preference, and the opportunity cost of capital suggests that a positive discount rate better reflects societal preferences. Similar to Stern (2006), we propose using a low discount rate (around the 0.5 percent mark) for the economic assessment of LD. This kind of value is particularly motivated by the long-term nature of LD processes and their impacts; it also reflects the need to assess the costs of LD from the perspective of society as a whole—now and in the future. The choice of discount rate for the valuation of environmental goods is essentially an ethical issue, not one of economic calculation only.

As stated earlier, the time preference of individuals, especially in developing countries, is often much higher than the value proposed here for the global analysis. For instance, for land degradation, a strong time preference can reflect constrained choices in land management due to poverty, leading to land-degrading rather than land-conserving practices, or due to the land user's high level of risk aversion. Typically, in poor rural areas, land degradation, poverty, and a high time preference are all interrelated, which can take the form of a vicious circle, as illustrated in Figure 1.4. As poverty increases, land users form ever-higher preferences for the present over the future, thereby mining the land resource further and increasing land degradation. The increasingly degraded land offers less ecosystem services and benefits upon which the land users (and others) depend, with the former sliding deeper into poverty and therefore favoring the present ever more.

The spiral effects depicted in Figure 1.4 can be escaped. An essential dynamic to stop is the ever-increasing level of poverty. There are several examples of schemes that target this dynamic through payment for ecosystem services, in particular when the land resource being held by the poor represents a type of international public good. More and more, public policies aim to make markets work better by integrating ecosystem service values, where possible, into price signals.

Figure 1.4—Illustration of the vicious circle connecting poverty, discount rates, and land degradation

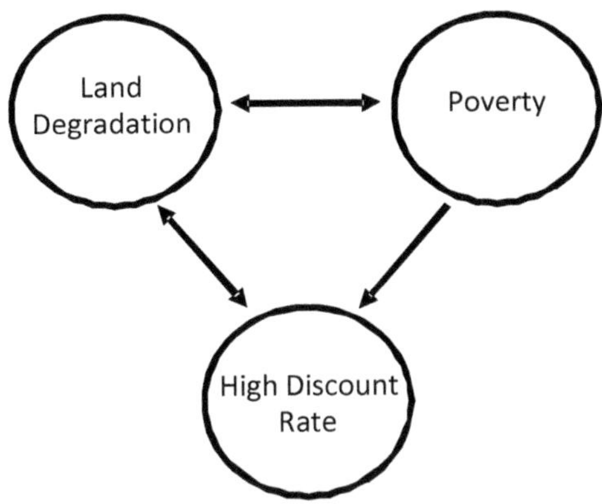

Source: Author's creation.

This technique helps internalize externalities in a way that helps pay for public benefits derived from ecosystem services of land. For example, addressing the lack of secure land tenure rights that contribute to land-degrading behavior is another way to internalize externalities. Fiscal instruments, such as taxes and subsidies, can help in that they consider the social costs and benefits instead of only private ones. Taxes might be imposed on unsustainable land management practices (or a subsidy may be offered for beneficial, land-conserving practices), so that the full environmental damage (benefit) to society is considered in the land use decision. Other innovative financial measures include incentive- and market-based approaches, such as the aforementioned payment schemes or trading (for example, emission trading or the clean development mechanism). The United Nations Reducing Emissions from Deforestation and Forest Degradation (UN-REDD) program is one example of a mechanism aiming to transfer payments to regions of the world with rich rain forest areas, with the goal of halting deforestation and mitigating its climate change impacts. However, the success of all these measures[7] and programs is embedded in their setup and in the rules governing them—the institutional

7 See Requier-Desjardins, Adhikari, and Sperlich (2010) for a more detailed review of instruments for the management of externalities suitable to prevent or mitigate land degradation and desertification.

arrangements mentioned in Figure 1.1. These institutional arrangements determine whether the payments provide the appropriate incentives to land users not to deplete the land resource. Payments for ecosystem services and the role of other incentive structures to decrease land degradation are discussed in Section 4.

The remainder of this report has the following structure: In Section 2, the report describes global assessments of land degradation and desertification, as well as the causes, processes, and consequences linked to LD. In Section 3, we take an economic perspective on LD, describing concepts of the economic valuation of the environment and reviewing numerous studies that carried out economic valuations of LD. Based on the review, we propose a methodological framework to assess the costs of action versus inaction based on marginal changes in LD. This framework draws on assessments at the microlevel (that is, where LD is actually observed) as well as at the aggregate level (that is, global models and scenarios). Section 4 highlights the key role of incentive structures and allows us to look in a disaggregated way at LD actors and their behavior within this realm of incentives. Section 5 reviews specific actions for improving degraded lands. Section 6 attempts to provide an assessment of LD based on case studies in the manner that we have proposed in the previous sections. It serves as illustration of our concept, no matter how preliminary or incomplete it might be compared to the global assessment that needs to be carried out. Section 7 proposes how to scale up to the global assessment, using the proposed methodology, in terms of the stakeholders who are to be included in the making of such a global assessment. Section 8 concludes.

2. Assessment of Land Degradation

2.1 Land Degradation

Land degradation is an extensive phenomenon influenced by natural and socioeconomic factors. As the problem is complex, the existing definitions of land degradation, the methods for its assessment, and the related actions are varied and sometimes conflicting. Although soil represents one of the key ingredients of land, there is a clear distinction between land and soil degradation; this distinction should be considered by researchers, land managers, and stakeholders.

Recognizing that the term *land* refers to more than just soil, the United Nations Convention to Combat Desertification defines *land* as "the terrestrial bio-productive system that comprises soil, vegetation, other biota, and the ecological and hydrological processes that operate within the system."(UNCCD, 1996, Part1, Article 1e) According to Vlek, Le, and Tamene (2008), the interaction of the land with its users is mainly what leads to any kind of land degradation, resulting in serious social problems, due to the change of the ensemble of the soil constituents, of the biotic components in and on it, and of its landscapes and climatic attributes. Because land use results in relevant services, such as food production and, more generally, support of livelihoods, land degradation directly affects social human benefits. Thus, interactions of natural processes, human activities, and social systems play a considerable role in land degradation (Safriel 2007).

Early definitions of *land degradation* refer to a decline in "the current and/or potential capability of soils to produce (quantitatively and/or qualitatively) goods and services" (FAO 1979). More recent definitions extend land degradation to spatial and time dimensions, as is reflected in the definition of the UNCCD, which defines *land degradation* in the context of its focus on drylands: the "reduction or loss in arid, semiarid, and dry subhumid areas, of the biological or economic productivity and complexity of rainfed cropland, irrigated cropland, or range, pasture, forest, and woodlands resulting from land uses or from a process or combination of processes, including processes arising from human activities and habitation patterns, such as

1. soil erosion caused by wind and water;
2. deterioration of the physical, chemical, and biological or economic properties of soil; and
3. long-term loss of natural vegetation." (UNCCD, 1996, Part1, Article 1f)

Per definition, land degradation can be caused by both human activities and natural events (Mainguet and da Silva 1998). With the impact of global climate change becoming ever-more evident, it is important to separate human-induced land degradation from that caused by climate change, over which land users have little or no control (Vlek, Le, and Tamene 2010). As a primarily human-induced environmental phenomenon (Johnson and Lewis 2007; Katyal and Vlek 2000) land degradation is therefore a social problem involving people at all stages not only as causative factors but also as victims (Blaikie and Brookfield 1987; Spooner 1987). Although, according to the UNCCD, land degradation is attributed to dryland ecosystems only, it is generally accepted that land degradation takes place in temperate climates as well (Akhtar-Schuster, Bigas, and Thomas 2010).

One of the main components of land degradation is desertification, which has been defined in many different ways by multiple disciplines. A first official definition was agreed upon in 1977 at the United Nations Conference on Desertification (UNCOD) in Nairobi (UNEP 1977): "Desertification is the diminution or destruction of the biological potential of land, and can lead ultimately to desertlike conditions. It is an aspect of the widespread deterioration of ecosystems and has diminished or destroyed the biological potential—that is, plant and animal production—for multiple-use purposes at a time when increased productivity is needed to support growing populations in quest of development"(Cited in Glantz and Orlovsky, 1983). According to Glantz and Orlovsky (1983), by the early 1980s, around 100 definitions of desertification were developed out of this initial definition, varying in the area of coverage, the causative factors, the anticipated impacts, and its reversibility (Glantz and Orlovsky 1983; Geist 2005). Katyal and Vlek (2000) provided a synthesis of definitions until 1994. Among these definitions, the official UNCCD definition describes desertification as "land degradation in arid, semiarid, and dry subhumid areas, resulting from various factors, including climatic variations and human activities" (UNCCD, 1996, Part1, Article 1a). Although this definition does not clearly distinguish between desertification and land degradation—though it explicitly includes climatic conditions as a causative element—it is now widely regarded to be the authoritative definition of *desertification*. Erroneously, many studies use the term *desertification* interchangeably with *land degradation* and vice versa.

Studies have focused on deforestation, overgrazing, salinization, soil erosion, and other visible forms of land degradation rather than on other less-visible forms of land degradation. Global cooperation in addressing land degradation issues emerged through United Nations (UN) conferences in the 1980s. Due to this cooperation, a large number of studies have assessed different forms of land degradation, each with differences in the accuracy of data. However, the new geographical information system (GIS) data, which facilitates the collection of large quantities of global time series data using satellite imagery, led to a significant increase in the accuracy of land degradation assessment. This technology has improved past methods, which relied heavily on expert opinion or extrapolation of localized estimation. Another weakness of these studies was their failure to determine the cost of degradation, an aspect that has made such studies less valuable to policymakers. For the few global studies that did assess the economic costs of land degradation, the methods and data used have been questioned. Despite their weaknesses, these earlier studies have shed light on the severity of land degradation and have helped raise the awareness of policymakers regarding the need to find strategies for combating land degradation. Later in this report, we review past studies, assess their strengths and weaknesses, and propose approaches that can be used to assess the costs of land degradation, as well as the costs and benefits of the prevention of land degradation and of the rehabilitation of degraded lands.

Generalized Map of the Status of Desertification in Arid Lands by the United Nations Environment Program for the United Nations Conference on Desertification (1977)

The first global map of estimated desertification was developed by the United Nations Food and Agriculture Association (FAO); the United Nations Educational, Scientific, and Cultural Organization (UNESCO); and the World Meteorological Organization (WMO) for the UNCOD. The map included data from 1977 and earlier (Dregne 1977; UNEP 1997; Thomas and Middleton 1994) and was produced without any georeferenced data. Three levels of desertification severity—moderate, high, and very high—were mapped to show areas prone to desertification. These degrees were based on an evaluation of climate conditions conducted by a limited number of consultants (Thomas and Middleton 1994). According to this approach, 35 percent of Earth's surface was affected by desertification in 1977. However, this study has been generally regarded to have overexaggerated the rate of land degradation. Recent studies using remote sensing and other data sources have shown that the extent of desertification is only between 10–20 percent, or between 6–12 million hectares (MA 2005a).

This first approach to mapping focused on desertification and drylands and therewith disregarded land degradation in general, while also excluding humid and subhumid areas. Moreover, georeferenced data was not used, which is essential for a real mapping approach.

Desertification of Arid Lands and Global Desertification Dimensions and Costs by Texas Tech University (1992)

A first study on the desertification of arid lands, conducted by Texas Tech University, used secondary data from 1983 and earlier, taken from 100 countries in six continental regions. Degradation attributes used include vegetation cover and composition, soil salinity and resulting crop yield reduction, and soil erosion. Results show that 30–100 percent of the mapping area was affected by severe land degradation, with all degradation categories making up 48 percent of land area. The resulting global and continental maps were published in Dregne (1983). However, like the UNCOD study, this study also suffers from poor data and information, as was acknowledged by Dregne (1983).

Texas Tech University did a second study on global desertification dimensions and costs (Dregne and Chou 1992), using combined data from 1992 and earlier and covering both soil and vegetation degradation. Together with data from the first study (Dregne 1983), they also used land use figures from FAO (1986) and primary data from additional field experiments. They chose as attributes those that affect economic plant yields, similar to those used for the 1983 assessment. Results indicate 70 percent land degradation of the assessed land. Although estimates are considerably better than the 1983 estimates, the database and information upon which calculations were made were still poor, with an upward bias.

Global Assessment of Human-Induced Soil Degradation by the International Soil Reference and Information Centre (1987–1990)

The Global Assessment of Human-Induced Land Degradation (GLASOD) study, which was carried out by the World Soil Information Center and commissioned by the United Nations Environment Program (UNEP) between 1987 and 1995, was the first major global assessment of soil degradation. The project represented "the basis of the most recent UN studies of global land degradation and desertification" (Thomas and Middleton 1996: 119). Within three years (by 1990), GLASOD developed a world map of human-induced soil degradation (Oldeman, Hakkeling, and Sombroek 1991b). The approach defines *soil degradation* as "human-induced phenomena, which lower the current and/or future capacity of the soil to support human life" (UNEP 1997b). The map was originally produced for the United Nations Conference on Environment and Development (UNCED) in 1992.

The GLASOD project was conducted using expert opinions of about 250 experts drawn from 21 regions of the world (Nachtergaele et al. 2010). Soil degradation was mapped by types and by country using the following degradation attributes: soil erosion by wind and water, chemical soil deterioration, and physical soil deterioration (for further explanation of types of degradation, see "Types of Land Degradation"). The authors then identified different degrees as light, moderate, severe, and very severe land degradation (Oldeman, Hakkeling, and Sombroek 1991a). The study observed that 38 percent of the global land area affected by human-induced land degradation was lightly degraded, 46 percent was moderately degraded and 15 percent was strongly degraded Oldeman 1998). Water erosion was identified as the most important form of soil degradation, followed by wind erosion, both of which accounted for about 84 percent of land area degraded (Table 2.1 and Figure 2.1). The most degraded area was Europe (25 percent), followed by Asia (18 percent) and Africa (16 percent); North America reported the smallest area degraded.

Table 2.1—GLASOD (1991) extent of human-induced soil degradation (in million hectares)

Type of land degradation	World	Asia	West Asia	Africa	LAC	North America	Australia and Pacific	Europe	% of total
Water erosion	1,094	440	84	227	169	60	83	115	55.70
Wind erosion	548	222	145	187	47	35	16	42	27.90
Nutrient depletion	135	15	6	45	72	-	+	3	6.87
Salinity	76	53	47	15	4	-	1	4	3.87
Contamination	22	2	+	+	+	-	-	19	1.12
Physical	79	12	4	18	13	1	2	36	4.02
Other	10	3	1	2	1	-	1	2	0.51
Total	1,964	747	287	494	306	96	103	218	100.00
% of total		38.03	14.61	25.15	15.58	4.89	5.24	11.10	

Source: Oldeman, Hakkeling, and Sombroek 1991a.

Notes: LAC=Latin American Countries; + means increasing land degradation; - means decreasing land degradation.

The GLASOD study was useful in formulating a number of global conventions and international and national land management development programs. GLASOD has also been quite useful in raising the attention of the extent and severity of soil degradation.

Figure 2.1—GLASOD (1991) global assessment of the status of human-induced soil degradation

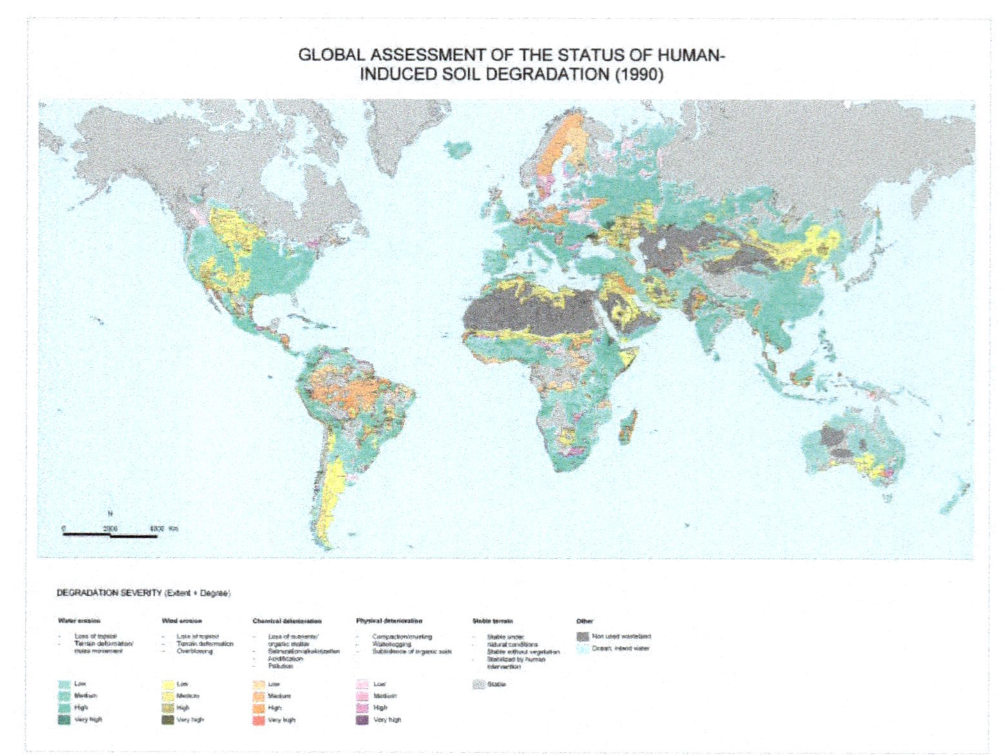

Source: Oldeman, Hakkeling, and Sombroek 1991a.

Sonneveld and Dent (2009) assessed the accuracy of the GLASOD results, using some case studies in Africa. The study concluded that the GLASOD assessments were only moderately consistent with the actual data used to verify it. The analysis of the relationship between agricultural productivity and the severity of soil degradation also showed a positive relationship between maize yield and the degradation index drawn from GLASOD, which was due to the use of fertilizer, improved seeds, and other yield-enhancing practices, excluding prevention of soil erosion. This counter-intuitive pattern shows the complex relationship between degradation and agricultural productivity.

The GLASOD study also focused on soil degradation largely related to soil erosion; only little is said about other forms of land degradation—namely, loss of vegetation (for example, deforestation) and biodiversity. As the GLASOD study was based on expert opinion, it is prone to subjective judgment that is not easy to reproduce. GLASOD focused only on the extent and severity of soil degradation, with no assessment of efforts to prevent degradation or to rehabilitate the results of soil degradation. The study also did not identify the causes of soil degradation and its costs. Due to these and other weaknesses, most of which the authors of the study identified, follow-up studies have been commissioned and are reviewed below.

In 1990, the first edition of the World Atlas of Desertification (WAD) by UNEP used the findings and products of the GLASOD approach for its atlas. The first edition was published in 1992, with the intention of depicting the status of desertification and land degradation. The second edition was published in 1997. In addition to the depiction of desertification, it contained methods to combat desertification, as well as responses to related issues, such as biodiversity, climate change, and the impact of socioeconomic determinants, such as population density (UNEP 1997b). Moreover, case studies in Africa and Asia were presented, which were based on the Assessment of Soil Degradation in Asia and Southeast Asia (ASSOD), the more regional approach of GLASOD.

Though based on soil degradation data provided by GLASOD, WAD claims to depict desertification instead. This underlines the need for common definitions of the terminology used.

World Overview of Conservation Approaches and Technologies

The World Overview of Conservation Approaches and Technologies (WOCAT) was the first global assessment of the prevention of land degradation mitigation. The WOCAT approach shows the wealth of local knowledge regarding soil and water conservation practices used by land users in different countries around the world. WOCAT started in 1992, with the intention of pointing out achievements that had been made to combat soil degradation (Schwilch, Liniger, and Van Lynden 2004). WOCAT was started in reaction to the GLASOD approach, in order to develop "a set tools to document, monitor, and evaluate" soil and water conservation (SWC) "know-how, to disseminate it around the globe, and to facilitate the exchange of experience" (Schwilch, Liniger, and Van Lynden 2004: 1). It was aimed at facilitating local and international exchange of knowledge by establishing the WOCAT network. A global map of SWC measures is still under construction. WOCAT also collected data on the costs of land management practices, which could allow a determination of the costs of taking action to prevent or mitigate land degradation. WOCAT conducted 42 case studies in 23 countries from six continents. However, the choice of the case study countries was not done in such a way that could allow extrapolation of the results to the rest of the world.

Global Desertification Tension Zones

The U.S. Department of Agriculture's Natural Resources Conservation Service (NRCS) developed global maps on *global desertification tension zones*, which depict areas vulnerable to desertification, wind and water erosion by natural preconditions, and human impact (Eswaran, Lal, and Reich 2001). Desertification tension zones were described with two systems: zones under low-input agricultural systems, where the "productive capacity of the land is stressed by mismanagement, generally by resource poor farmers," and high-input agricultural systems, where tension zones "arise due to excessive use of agrochemicals, uncontrolled use of irrigation, and monoclonal plantations with minimal genetic diversity" (Eswaran, Reich, and Beinroth 1998: 1). Four degrees of severity (low, moderate, high, and very high), reminiscent to the GLASOD approach, show the degree of an area's vulnerability to desertification or land degradation (Figure 2.2). According to the definition of *desertification* from UNEP, only arid, semiarid, and subhumid areas were integrated in this approach, excluding subhumid and humid, except for water erosion, which also affects humid areas. The purpose of the NRCS study was to identify and locate "desertification tension zones [...] where the potential decline in land quality is so severe as to trigger a whole range of negative socioeconomic conditions that could threaten political stability, sustainability, and the general quality of life" (Eswaran, Reich, and Beinroth 1998: 1). The approach of NRCS used several data input, such as the FAO/UNESCO Soil Map of the World (1:5,000,000) and a pedoclimate map that was compiled by a climate database from records for about 25,000 stations globally, using soil moisture and temperature regimes that were superimposed on the soil map via GIS. The map of vulnerability to desertification was superimposed with an interpolated population density map from the Center for International Earth Science Information Network (CIESIN) from 1995. The output was a map showing areas subjected to the risk of human-induced desertification (Figure 2.3) In NRCS, the socioeconomic determinants that cause an area to be vulnerable to desertification were only approximated by population densities. However, there may be regional differences that influence whether high population densities trigger land degradation or avoid it (see, for example, Vlek, Le, and Tamene 2010). When interpreting this map, it must be kept in mind that it shows "vulnerability to desertification" and not the current state of desertification.

Figure 2.2—Desertification vulnerability

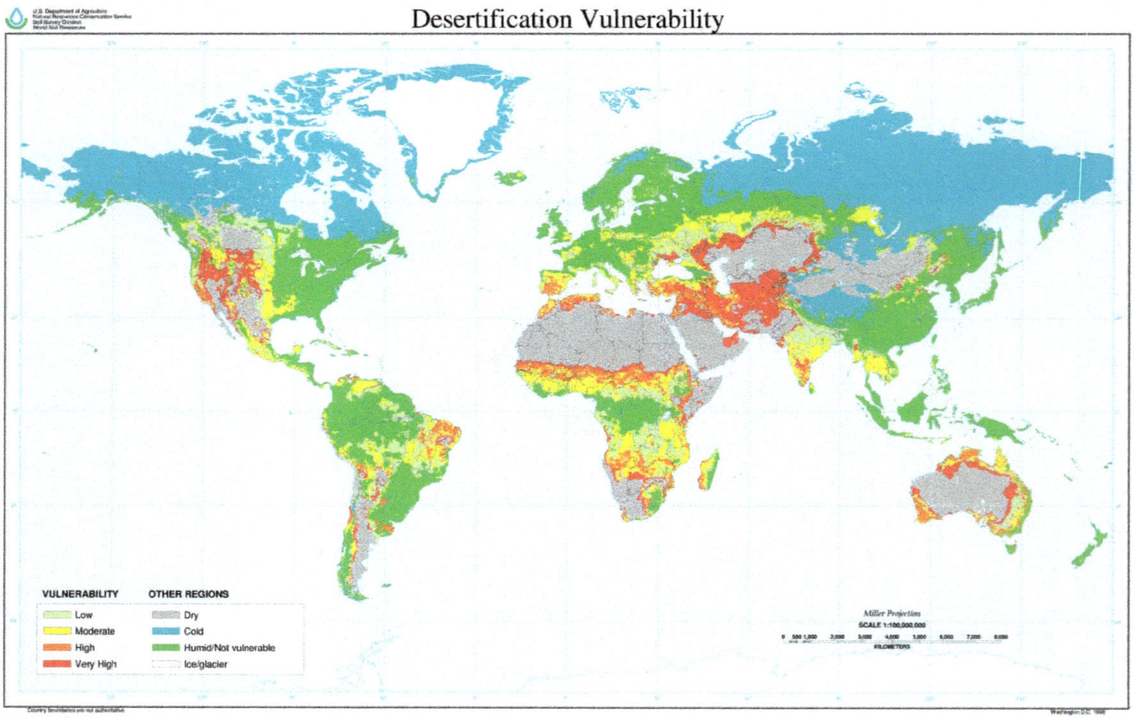

Source: USDA-NRCS 1998.

Figure 2.3—Areas vulnerable to human-induced desertification

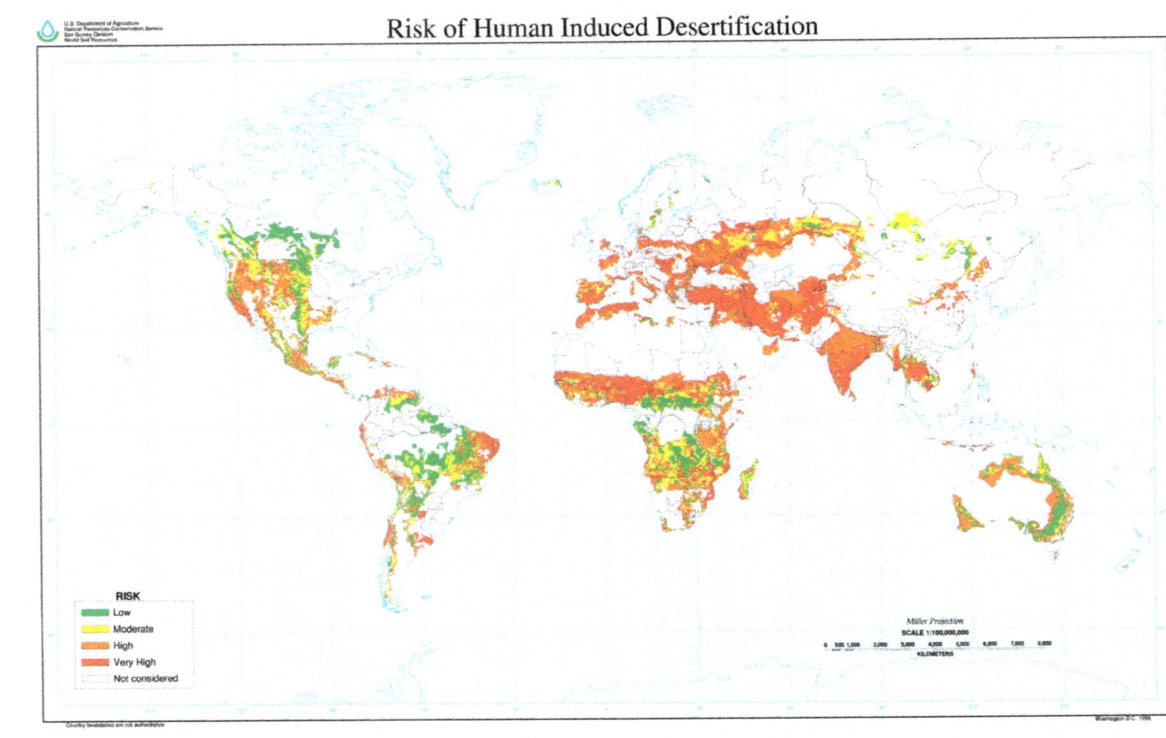

Source: USDA-NRCS 1999.

Land Degradation Assessment in Drylands/GLADA

Taking advantage of GIS and remote-sensing technology, FAO initiated the Land Degradation Assessment in Drylands (LADA), which has produced assessments on the causes, status, and impact of land degradation in the drylands. The study uses existing maps and databases but also incorporates new information from satellite imagery. The Global Land Degradation Assessment (GLADA) built on LADA and assessed both land degradation and land improvement. GLADA's global coverage also included humid and subhumid areas. Unlike GLASOD, GLADA study used satellite data rather than expert opinion.

GLADA defined *land degradation* as a "long-term decline in ecosystem function and productivity" (Bai et al. 2008b: 1). Net primary productivity (NPP), or the rate of carbon dioxide fixation by vegetation less losses through respiration, was used to measure ecosystem productivity (Bai and Dent 2008). Using remote-sensing data, GLADA could measure both land degradation and land improvement by the change in NPP. GLADA also employed a much higher spatial resolution (8 kilometers) than what was used before.

According to Oldemann (2002), LADA is based on two principles:

- The assessment of nature and the quantification of the extent and severity of land degradation and its impact on the environment and human society;
- The building of capabilities to design and plan interventions to mitigate land degradation

Vegetation monitoring can be used as a proxy for land degradation assessment by measuring NPP. First, a 23-year period (1981–2003) of fortnightly Normalized Differenced Vegetation Index (NDVI) dataset was used for GLADA, thus representing the global component of LADA. This period was later extended to a 26-year period (1981–2006). The indicators monitored with remote-sensing imagery were NPP, rainfall use efficiency (RUE), aridity index (AI), and rainfall variability and erosion (see Box 2.1 and Appendix B, Figure B.2). How the different indicators are related or can be combined is still debated among scientists. The indicators were interpreted based on a global land use system map, which also adds socioeconomic determinants.[8]

8 Change detection describes a remote-sensing technique in which two or more satellite images can be detracted from each other to detect changes in the land surface.

Box 2.1—Tools and indicators of land degradation, land improvement, and of their socioeconomic drivers

Monitoring Tools

Geographical information systems (GIS) allow for the analysis, transformation, and modeling of geographic data. Metadata can be derived from remote-sensing imagery as well as from ground data surveys (Venkataratnam and Sankar 1996). Similar to modeling, this tool is usable on the global, national, and local scale. The soil and terrain (SOTER) digital database of the world project, which uses GIS and was initiated in 1986 by the International Society of Soil Science (ISSS), provides global soil information and can be useful in obtaining raw data to start the assessment at regional scales (Dobos, Daroussin, and Montanrella 2005).

Remote Sensing

Since the 1980s, remote sensing has played a large role in the acquisition of spatial–temporal data. Remote sensing is widespread in the global and local assessment of land degradation, as it allows data collection without using in situ field data collection, which is time consuming and expensive. Remote sensing is divided into two types: (1) *Passive remote sensing systems* record natural electromagnetic radiation, mostly solar radiation, reflected at bodies' surfaces; and (2) *active remote sensing* uses its own source of energy or illumination, which means that an energy source irradiates a body and collects the reflected radiation afterward by itself (Lillesand, Kiefer, and Chipman 2004). GLASOD, GLADA/LADA, and GLADIS use passive remote sensing systems, such as optical satellite imagery, for monitoring physical determinants of land degradation.

Radio Detection and Ranging (RADAR)

Since the 1990s, this active remote-sensing tool has been used for environmental monitoring purposes. Radar uses microwaves with band lengths between 3 and 25 centimeters (Jensen 2000). Because of its long wavelengths, radar is not influenced by clouds or other atmospheric weather conditions, which makes it especially useful for land degradation assessment in tropical regions, where clouds negatively influence passive remote sensing systems in general. By using microwaves, we can get information on soil moisture and, therewith, on soil conditions and roughness as compaction or aridification (Leberl 1990). Although this information is needed for land degradation assessment, only a few approaches to land degradation or desertification have already used this method (De Jong and Epema 2001).

High-Resolution (HR) to Very High-Resolution (VHR) Satellite Imagery

This tool enables the observation of land surface changes within a certain period. Via change detection, significant differences in soil or vegetation surfaces can be identified from two or more satellite images of different times. In the past two decades, geospatial analytical methods and tools have rapidly advanced, enabling more rigorous yet cheaper analyses of land degradation and improvement (Buenemann et al. 2011). We review some key indicators used for land degradation and improvement assessments. Land degradation and improvement that affect provisioning services have been measured in terms of land cover, biological productivity, and water quality and quantity.

Box 2.1—Continued

Indicators of Land Degradation and Improvement

Normalized Differenced Vegetation Index (NDVI)

Vegetation cover is considered a key indicator of the status of land. The most commonly used vegetation index is the Normalized Differenced Vegetation Index (see, for example, Bai et al. 2008b; Vlek, Le, and Tamene 2008). The NDVI is calculated from the remotely sensed reflection of vegetation surfaces of visible (red) and near-infrared light over their sum and gives information on the density, condition, and health of photosynthetically active vegetation (Helldén and Tottrup 2008). Accounting for climate, soils, terrain, and land use, a deviation from the norm is an indication of either land degradation or improvement. The higher the NDVI value, the higher the photosynthetic activity. NDVI has increasingly been used to assess a variety of ecosystem services. However, NDVI only measures vegetation and may therefore conceal some forms of degradation. For example, weed encroachment and invasive alien species are counted as land degradation, but NDVI captures such trends as land improvement. Carbon fertilization increases vegetation even on degraded lands, thus concealing degradation (Vlek, Le, and Tamene 2010).

Net Primary Productivity (NPP)

NPP is the rate of carbon dioxide fixation by vegetation without losses through vegetation respiration (Bai et al. 2008b). Techniques to measure NPP by Earth-observing satellite data were implemented in the mid-1980s (Prince 1991). The NPP can be used to define the carbon balance for terrestrial surfaces (Justice et al. 1998). NPP and NDVI are related and therefore share the same weaknesses.

A negative trend in NPP does not necessarily indicate land degradation; and, likewise, a positive trend does not directly indicate land improvement. False alarms are also common when using NPP.

Rainfall Use Efficiency (RUE)

RUE is calculated by determining the ratio of NDVI or NPP to annual precipitation. Rainfall is a determining factor of vegetation and, thus, of production. RUE assesses the dependence of vegetation and rainfall as a "key indicator for functioning of semiarid ecosystems" (Le Houérou 1984: 752). In general, RUE of degraded areas is lower than RUE of nondegraded areas (Safriel 2007). Because vegetation reacts in a short time to natural variations, RUE needs to be monitored over the long term to exclude false alarms.

Residual Trend Analysis (RESTREND)

This analysis shows the relationship of NPP (a math formula representing the sum of NDVI (ΣNDVI)) to rainfall and measures the negative trends in the difference between the observed ΣNDVI and the predicted ΣNDVI by rainfall. Nonetheless, according to Wessels et al. (2007), the causes of a negative or positive RESTREND have to be assessed by local investigation, because potential land degradation can be identified only on a regional scale.

Box 2.1—Continued

Critique of Indicators

NDVI and NPP are often referred to as "lumped" parameters, because they reveal aggregate and complex information. As noted above, NDVI does not allow for differentiation between various forms of vegetation and their health status, making it vulnerable to false alarms such as bush encroachment, which is an indicator for a decrease in soil quality and therewith land degradation, but depicts higher NDVI results in relation to the actual situation. It gives the idea of a more dense and therewith productive land, whereas it actually reflects land degradation. In addition, land use change, or the conversion of agricultural areas into urban areas, cannot be identified with NDVI, leading to low or negative NDVI values even though there is no land degradation. Within arid to semiarid ecosystems, NDVI can be misleading because of canopy background variations, which restricts the accurate assessment of the vegetative cover, especially since soil spectral variations are far greater in arid and semiarid regions than in the more humid, organic-enriched grasslands and forests (Huete, Justice, and Liu 1994). Canopy background considerations are important not only in sparsely vegetated areas but also in woodlands, savannas, and open forest stands. Improved variants to the NDVI equation include a soil-adjustment factor or the blue band for atmospheric normalization.[9] However, NDVI is the only index used at global scales, because it can be easily measured without additional information (on soil and so on) and because long-term global data are freely available. Although correcting NDVI with global socioeconomic data can help ameliorate some false alarms (see next section), there is still a need for a more complete biophysical understanding of the indicators to help validate them as well as to provide estimates of their accuracy and uncertainty (Huete, Justice, and Liu 1994). RUE, which is based on rainfall, suffers from several weaknesses: Rainfall is not necessarily a limiting factor of vegetation growth, especially in humid areas and forests, and it takes into consideration neither the intensity nor variability of rainfall.

Monitoring and Assessing Socioeconomic Indicators

In terms of socioeconomic indicators, mostly population data, such as population density, have been taken into account (by, for example, USDA-NRCS and GLADIS). According to the Sede Boqer approach,[10] in 2006 the increase of population densities in particular represents the main human impact triggering the reduction of the productivity in drylands (Safriel 2009). Several national approaches, such as the Land Degradation Monitoring of Namibia (Klintenberg and Seely 2004), as well as livestock pressure (the number and density of livestock), took into account the ongoing problem in rural areas of arid regions.

A newer technique for mapping socioeconomic determinants could be remote sensing, which is already used for biophysical land degradation and improvement. There is a need for accurate maps on the global scale that depict the extent, location, and size of human settlement; such maps could help policymakers, as well as researchers, understand the collective impact of human settlement development and thus anticipate further growth.

---

9 The SAVI (Soil-Adjusted Vegetation Index) was designed to remove much of the ground contamination associated with soil-brightness variations (Huete 1988).

10 A new approach on sustainable development in drylands, developed in an international conference that took place at the end of the International Year of Deserts and Desertification 2006, hosted by the Blaustein Institute for Desert Research in Sede Boqer, Israel.

Box 2.1—Continued

Urban areas are those areas in which human activity is the highest, accounting for 50 percent of the world's population and 70–90 percent of economic activities (Schneider, Friedl, and Potere 2010). The observation of urban areas (for example, their development in spatial distribution) could provide an opportunity to identify false alarms within NDVI-based maps that occur due to land use and change. Cultivated land or land in general that is transferred into urban areas could be identified and corrected. A new global map of urban areas, implemented by the Moderate Resolution Imaging Spectroradiometer (MODIS), defines urban areas as places that are dominated by built environments, including nonvegetated and human-constructed elements, as all human-made constructions (Schneider, Friedl, and Potere 2010). In 2009, the Socioeconomic Data and Application Center (SEDAC), hosted by CIESIN, developed the Global Rural–Urban Mapping Project (GRUMP) in cooperation with the U.S. National Aeronautics and Space Administration (NASA). GRUMP provides information on rural and urban areas along with the spatial distribution of human population in a common georeferenced framework derived from satellite and statistical data (SEDAC 2010). GRUMP has been used to describe the distribution of human settlements in low-elevation coastal zones (LECZs) around the world and has also been used by MA.

The study of infrastructure analysis has been used to analyze the impact of roads on vegetation. Road infrastructure is captured by satellite imagery, and a number of satellites capture road data. For example, global assessment of roads was done by Nelson (2008) and has since been used extensively in other studies. In addition, a number of organizations offer free road infrastructure data: the United Nations Spatial Data Infrastructure, the United Nations Global Alliance for Information and Communication Technologies and Development Open Access to and Application of Scientific Data in Developing Countries, OpenStreetMap, Global Roads Open Access Data Set, and others.

Six pilot countries represent the national-level LADA: Argentina, Cuba, China, Senegal, South Africa, and Tunisia. These countries represent important drylands on different continents (Tengberg and Torheim 2007). Within each of the six pilot countries, two to six local areas were selected for detailed assessment (FAO 2008). Six assessment methodologies - expert opinion, remote sensing, field monitoring, productivity changes, farm-level studies, and modeling - were chosen to get a broad view of the process at the national and local levels (Van Lynden and Kuhlmann 2002). The local assessment of LADA was conducted in collaboration with the University of East Anglia and WOCAT, which consider both biophysical and socioeconomic indicators (Bunning and Ndiaye 2009).

The remotely sensed NDVI is used as a proxy for changes in ecosystem productivity, accounting for climate, soils, terrain, and land use. Thus, a deviation from the norm is an indicator of land degradation or improvement. NDVI is strongly correlated with NPP (Figure 2.4), as shown recently by Vlek, Le, and Tamene (2008) and Bai et al. (2008b).[11]

11 The conversion from NDVI to NPP is reasonably easy in drier climes but seems to become unreliable in humid regions (P. L. Vlek, pers. comm., 2011).

Figure 2.4—Loss of annual NPP, GLADA, 1981–2003

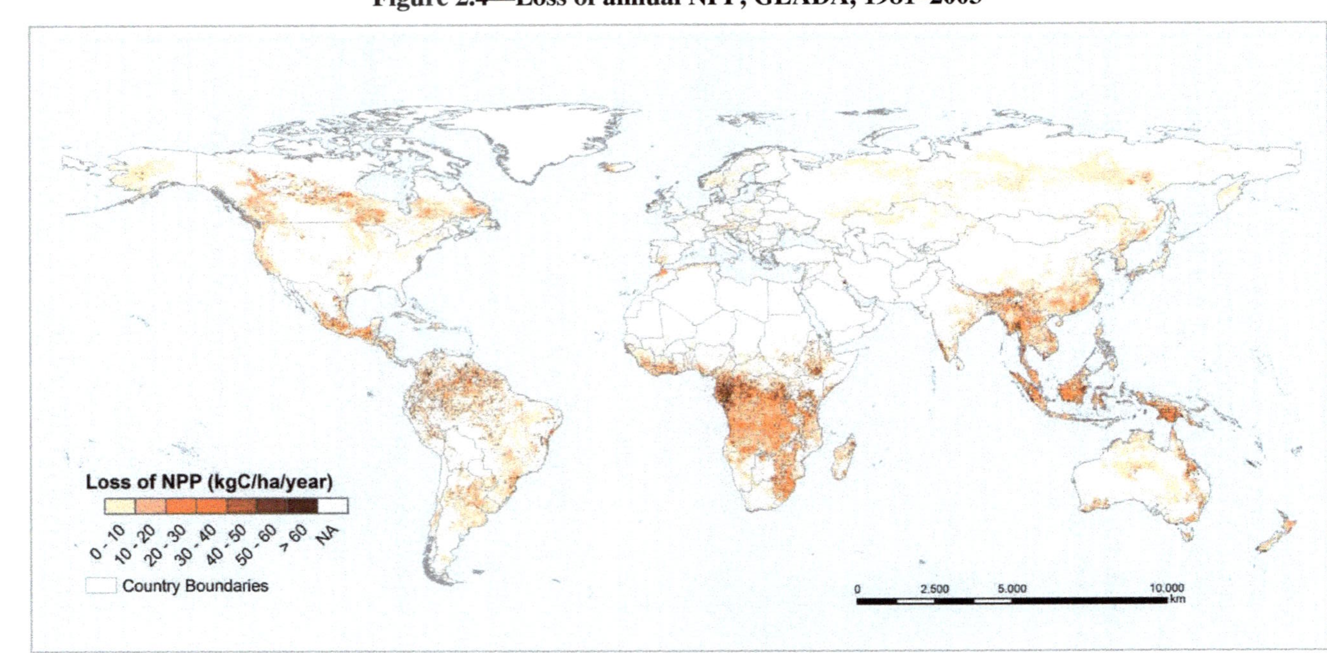

Source: ISRIC – World Soil Information, 2008.

Although first studies for South Africa and Kenya have been published, the approach of combining two indexes is still controversial (Vlek, Le, and Tamene 2008). About 80 percent of the degraded area in 1981–2003 occurred in the humid area (Figure 2.5). Because this percentage is surprisingly high, it was criticized by Wessels (2009), who indicated that RUE is a poor indicator in humid areas, which means this figure may not be reliable.

Figure 2.5—Degraded area as a percentage of total global degraded land area across agroclimatic zones, GLADA, 1981–2003

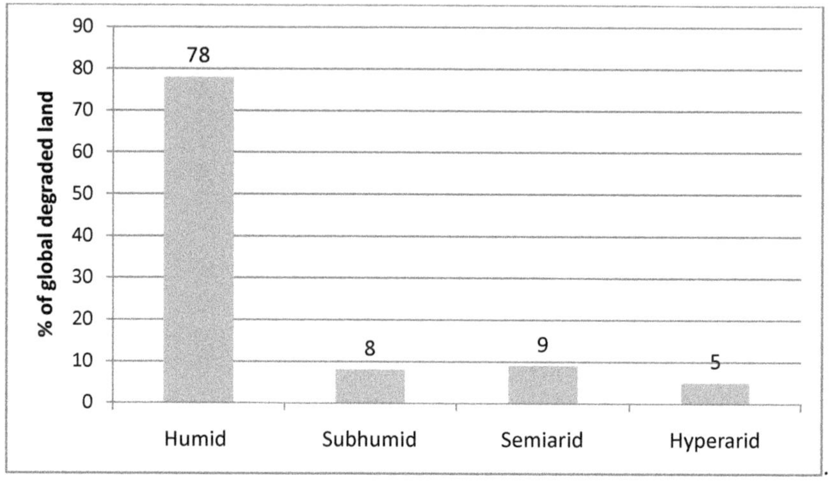

Source: Bai et al. 2008b.

The area in which land degradation was most severe was Africa south of the equator, which accounted for 13 percent of the global land area and 18 percent of NPP loss (Figure 2.6). The region of Indochina, Myanmar, and Indonesia was the second most severely degraded, accounting for 6 percent of global land area and 14 percent of lost NPP. The relationship between aridity and land degradation, measured as a decrease in NDVI, was negative at global level, suggesting that the extent and severity of land degradation was more severe in humid and subhumid areas than in semiarid, arid, and superarid areas. This finding is contrary to conventional wisdom, which states that drylands are more degraded than humid areas. Unfortunately, Bai et al. (2008b) did not offer an explanation for this finding. Wessels (2009) argued that the negative trends might rather be due to management practices, such as logging and crop rotation, than to land degradation; hence, the results require analysis of the causes of land degradation. The GLADA study examined the relationship between land degradation and poverty and population density.

Figure 2.6—Areas most affected by land degradation, GLADA, 1981–2003

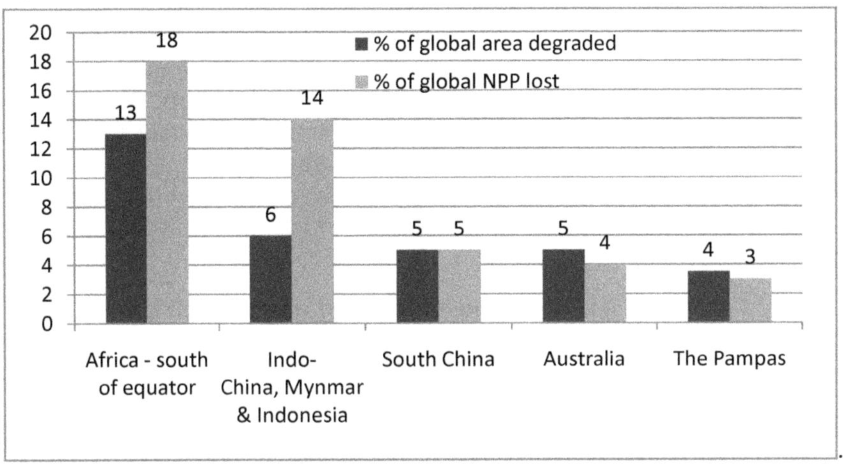

Source: Compiled from Bai et al. 2008b.

Although causal relationships cannot be derived from this approach, the results challenge conventional wisdom, pointing out greater land degradation in areas with high population density, though it must be stressed that land degradation was measured as changes in NDVI or NPP indicators. The GLADA study observed a negative relationship between population density and land degradation, supporting studies that observed the phenomenon of "more people less erosion" (Tiffen, Mortimore, and Gichuki 1994; Vlek, Le, and Tamene 2008). Vlek, Le, and Tamene (2008) offered as an explanation that these areas may constitute marginal lands with low carrying capacity, which can easily be overpopulated. The GLADA study also observed a positive correlation between poverty measured as a proportion of mortality rate of children under five years old and land degradation, supporting other studies that observed a vicious cycle of poverty and land degradation (Way 2006). Tree planting in Europe and North America and land reclamation in northern China increased the NDVI. Woodland and bush encroachment into rangeland and farmland also increased, contributing to a positive NDVI trend. Overall, land area improvement accounts for 16 percent, with rangelands contributing 43 percent of the improvement and with forests and crop areas contributing 23 percent and 18 percent, respectively. However, the increase of NDVI was not attributed to atmospheric fertilization, which describes the rising carbon dioxide levels of the atmosphere and the corresponding vegetation growth and which might, hence, be overestimated.

Weaknesses of the GLADA study, as acknowledged by the authors, include the usage of still-coarse data of 8 kilometers. The validation of the global assessments based on field-level observations in several countries often contradicted the GLADA results (for example, in South Africa, only 50 percent of the global predictions was correct). NDVI as an indicator of land degradation has shortcomings, as vegetation depends on several factors—not just the degradation status of the land. Wessels (2009) criticized the summation of NDVI over calendar years instead of summing over the vegetation period. The GLADA study also shows degradation in areas where there is sparse population density. For example, Gabon and Congo show the most severe land degradation (Figure 2.7), but population density in these two countries is among the lowest in Sub-Saharan Africa.[12]

Figure 2.7—Annual loss of NPP in eastern and southern Africa, 1981–2003

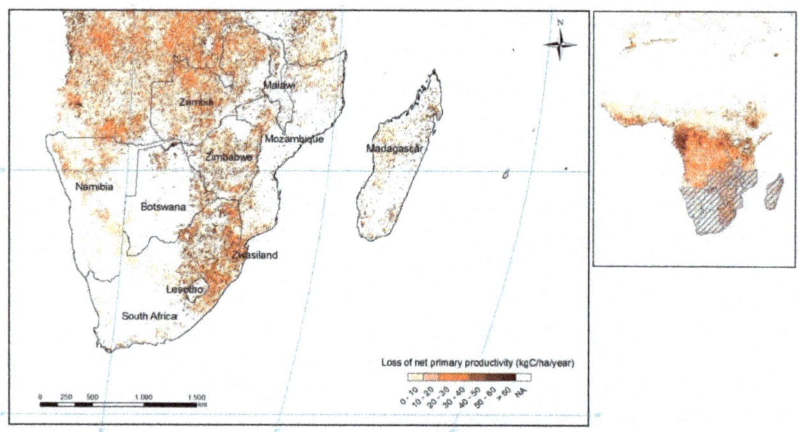

Source: ISRIC – World Soil Information, 2008

In GLADA, only simple correlation analysis with land degradation was applied. For example, the positive association between population density and land degradation does not control for many other factors that could simultaneously affect land degradation. The study also did not attempt to analyze other factors that could affect land degradation, even though it generated pixel-level socioeconomic data that could be used to analyze the effect of socioeconomic drivers of land degradation; such analysis would help in understanding the

12 The population densities of Gabon and Congo are, respectively, 6 persons per square kilometer and 12 persons per square kilometer, while the average population density in Sub-Saharan Africa is 35 people per square kilometer.

required policies and strategies for addressing land degradation. Furthermore, the study did not evaluate the cost of land degradation and the benefits of preventing land degradation and rehabilitating degraded lands. The simple relationship of yield and land degradation is likely to be masked by other practices used to increase productivity in degraded areas. For example, we examined the yield trend in African countries that experienced the most severe loss of NPP reported by GLADA (Figure 2.7) and observed an upward trend of major cereals in eastern and southern Africa and in Cameroon (Figures 2.8–2.10). The major reason behind the increasing yields in eastern and southern Africa and Cameroon is an increase in the use of improved seeds and fertilizer. Excluding South Africa and Zimbabwe in southern Africa, nitrogen application rates increased by about 10 percent from 2002–2004 to 2005–2006 in countries that showed severe land degradation (Table 2.2). Investments in infrastructure and other rural development programs have also been key to Africa's fast growth in the past decade (Foster and Briceño-Garmendia 2010). The results imply that the two-way relationship between land degradation and agronomic yield is likely to give misleading conclusions. Better results would be obtained if all major determinants of yield were included in models in order to determine the actual impact of land degradation on crop yield. Moreover, neglecting any valuation of crops or crop utilization implies neglect of economic effects.

Figure 2.8—Yield trend of major cereals in southern Africa, 1981–2009

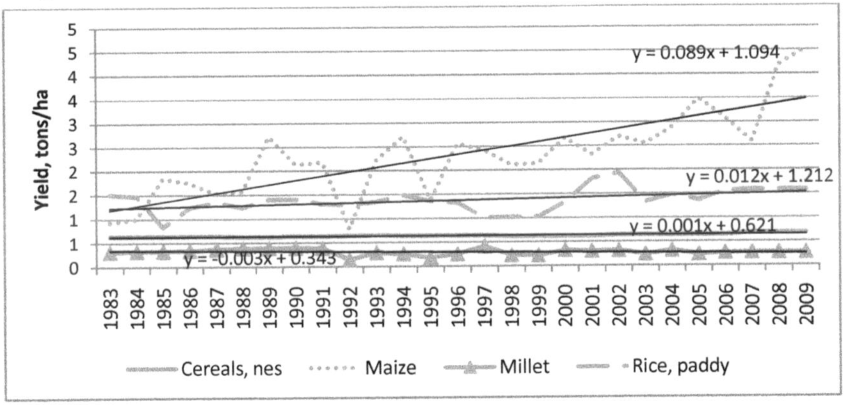

Note: nes = not elsewhere specified.

Source: FAOSTAT.

Figure 2.9—Yield trend of major cereals in eastern Africa, 1981–2009

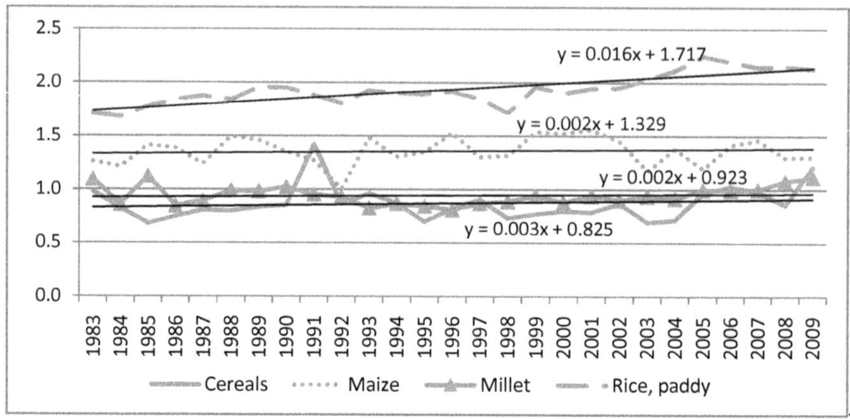

Source: FAOSTAT.

Figure 2.10—Cereal yield trend in Cameroon, 1981–2009

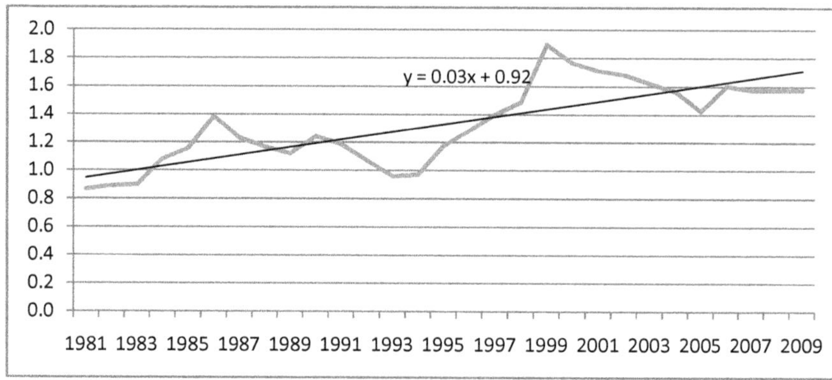

Source: FAOSTAT.

Table 2.2—Nitrogen application rates in selected eastern and southern African countries

	2002–2004	2005–2006	Change (%)
	kgN/ha	kgN/ha	
Angola	1.10	1.44	30.2
Malawi	21.24	23.90	12.5
Mozambique	3.24	2.44	-24.7
Namibia	2.32	1.77	-23.8
Zambia	21.49	21.75	1.2
Uganda	0.54	0.46	-15.3
Tanzania	2.61	4.88	86.8
Kenya	11.65	13.11	12.5
Cameroon	3.13	3.45	10.3
Average	7.48	8.1	10.0

Note: kgN/ha = kilograms of nitrogen per hectare.

Source: FAOSTAT.

Millennium Ecosystem Assessment, 2001–2005

The Millennium Ecosystem Assessment (MA) was conducted from 2001 to 2005 in order to assess current trends in ecosystems and human well-being. The study was part of a UN effort to evaluate the impact of changes of ecosystem services on human well-being (MA 2005b; Lepers et al. 2005). The MA report, published in 2005, was based on 14 global, regional, and subregional studies, including remote sensing and other data sources, with georeferenced results compiled into a map with a spatial resolution of 10 kilometers by 10 kilometers. The period from 1980 to 2000 was not provided in all datasets; therefore, no precise information on degradation within a certain period is given (Lepers et al. 2005).

Unlike GLASOD and GLADA, MA covered a broad range of ecosystem services (see Box 1.1). MA also attempted to assess the drivers of changes to ecosystem services and their impact on human well-being, as well as providing rich ecosystem data. The MA used GLASOD, the Convention on Biological Diversity, and other available data and studies to assess land degradation, which helped to enrich its assessment by taking advantage of past studies. These aspects make the MA study informative in policy advice.

The MA study points out that the drivers of changes in ecosystem services are multiple and related in a complex and interactive manner. These drivers of changes mediate and lead to trade-offs and synergetic associations. However, some general patterns could have been established as drivers of major changes. Conversion of land use types to agriculture has led to loss of biodiversity. The

MA study also suffers from some weaknesses, because it does not examine the costs of land degradation or the costs and benefits of the prevention of loss of ecosystem services or the rehabilitation of degraded ecosystem services.

Land Degradation in Sub-Saharan Africa

The studies by Vlek, Le, and Tamene (2008, 2010) analyze land degradation in Sub-Saharan Africa, using NDVI for the same period as GLADA (1982–2003) in order to approximate NPP. The NDVI data were based on mean NDVI per year and per month. RUE was not used in this study; instead, the correlation between rainfall and NDVI was assessed to differentiate between climate-driven and human-induced productivity changes. Figure 2.11 shows the declining or improving trends in vegetation over the 22 years. As a result, 10 percent of Sub-Saharan Africa (2.13 million square kilometers) showed a significant decline in NDVI, presumably caused by land degradation. Furthermore, the authors argued that atmospheric fertilization may be responsible for the extensive greening found in the analysis of NDVI trends (see Figure 2.12), masking underlying land-degrading processes. They selected areas with low population density, insignificant rainfall–NDVI correlation, and positive NDVI trends in order to estimate the rate of NPP improvement attributable to atmospheric fertilization. The study observed that in 17 percent of the area of Sub-Saharan Africa, land degradation was more than compensated for by atmospheric fertilization. Taking atmospheric fertilization into account, the identified degraded areas fit reasonably well with GLASOD results but do not fit the GLADA results based on RUE-adjusted NDVI (Figure 2.4). Though Bai et al. (2008b) identified a similar figure (26 percent of Sub-Saharan Africa's land area) experiencing land degradation, a geographical overlap of the affected land areas and population density is missing. To analyze the role of population as a factor influencing land degradation, the authors used three classes of population density: low, high, and very high. Consistent with Bai et al (2008b), Vlek, Le, and Tamene (2010) found low population densities in areas most affected by degradation. These areas may constitute marginal or fragile lands with limited carrying capacity. In addition, in some areas, high population densities are associated with land degradation on probably more fertile lands.
Examples of such areas are the densely populated areas in western Africa—especially the humid southwestern areas. In those areas, degradation problems could be addressed by improved access to fertilizer and erosion control measures. The authors also used FAO soil classes to assess whether soil and terrain constraints affect NDVI decline; they observed that 30 percent of the degraded areas relate to unsuitable agricultural soils. In addition, information on land use can help explain the human impact by the land use type of the degraded land. Finally, the authors analyzed the pressure of anthropogenic activities on

land using Human Appropriation of Net Primary Production (HANPP), which is the amount of NPP used by humans (for example, for harvested crops). The higher the HANPP (expressed as a percentage of NPP), the greater the human consumption or appropriation.

Figure 2.11—Long-term degradation of green biomass (1982-2003) of SSA

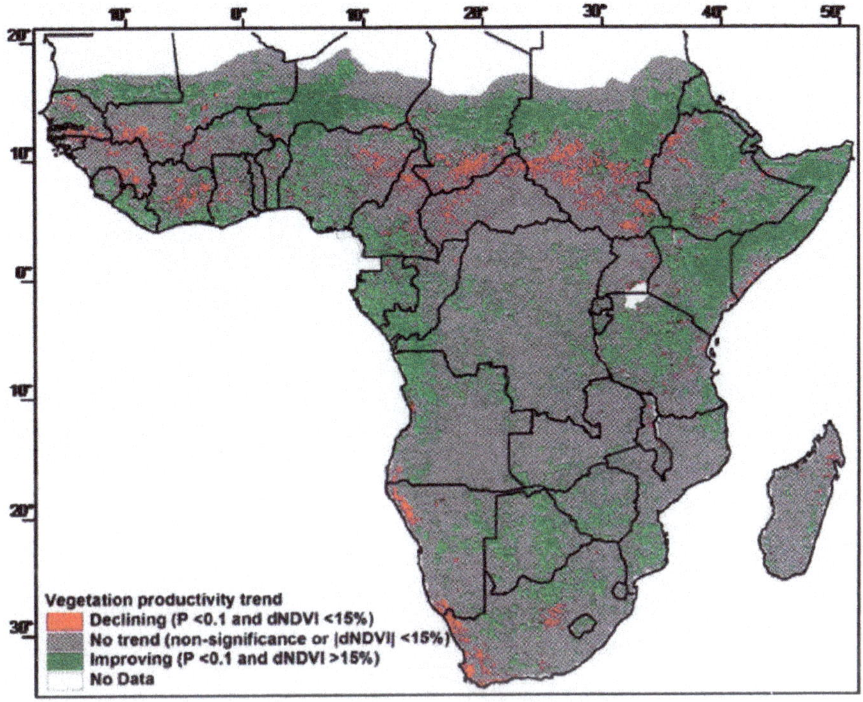

Source: Vlek, Le, and Tamene 2010.

Note: Confidence levels of 90% and relative productivity change of at least 15% over the last 22 years.

For Sub-Saharan Africa, the average HANPP suggests a light human impact, though there is wide variability across countries. High values of HANPP are associated with areas known for food insecurity and may also reveal areas with future insecurity. This finding is consistent with Bai et al. (2008b), who observed a positive correlation between land degradation and poverty. However, Bai et al (2008b) admitted that this study can only be seen as a way to locate global hot spots that appear to be threatened by human-induced land degradation; further verification and analysis in the field would be needed. The coarse resolution of 8 kilometers may also hide improvement or degradation; finer resolution is required to give better estimates. This study is also only limited to Sub-Saharan Africa; future studies should provide analysis in other regions of the world.

Figure 2.12—Areas affected by human-induced land degradation measured by a declining NDVI (Change in NDVI from 1982–2003 with a three-year base- and endline)

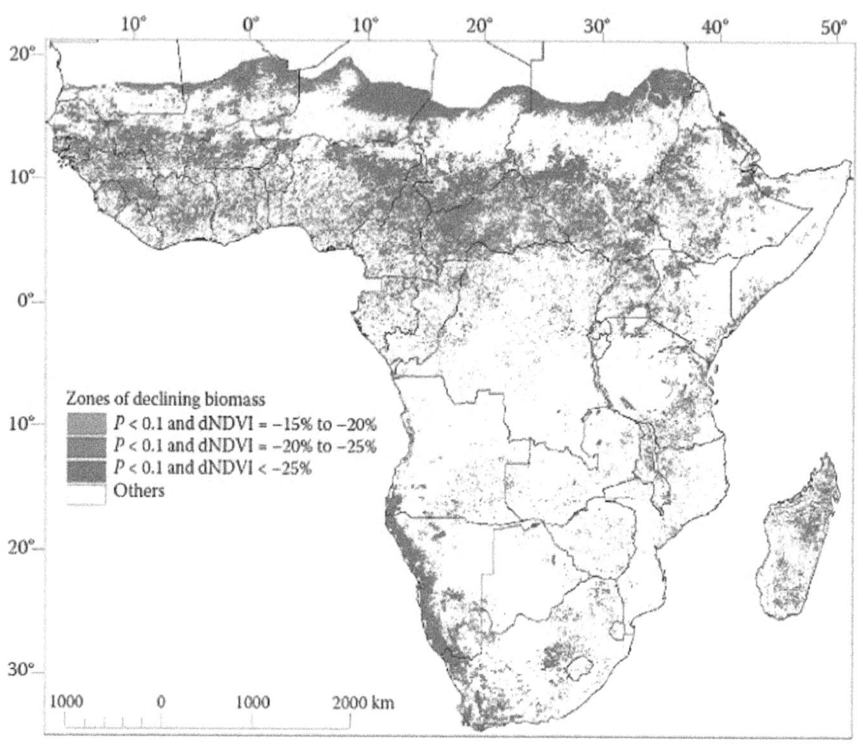

Source: Vlek, Le, and Tamene 2010.

Global Land Degradation Information System (2010)

The Global Land Degradation Information System (GLADIS), published as a Beta Version in June 2010, will be the final product of LADA. It is based on the ecosystem approach and combines pre-existing and newly developed global databases to inform decisionmakers on all aspects of land degradation at this scale (Nachtergaele et al. 2010). GLADIS provides a range of maps on the status

and trends of the main ecosystem services, supplemented with maps and databases on the physical and socioeconomic parameters (Nachtergaele et al. 2010). Land degradation was perceived as a complex process, the assessment of which must address more than only biophysical indicators with more than only one method. GLADA mainly used remote sensing—in particular, NDVI—for land degradation monitoring on a global scale. GLADIS is based on six axes for biomass, soil, water, biodiversity, economics, social indicators, and cultural indicators. It aims to capture the present status of land resources, as well as the degradation processes acting on them.
According to Nachtergaele et al. (2010), the status of the ecosystem's provisioning capacity is as important as its decline. Therefore, the degradation and improvement of ecosystem services are presented in GLADIS.

The WebGIS of GLADIS[13] is divided into four sections (Land Use Systems, Database, Analysis and Land Degradation Index). The Land Use Systems of the World map (see Appendix B, Figure B.1) contains about 40 different classes that give information about land cover,[14] livestock density, and management (irrigation, protected, unmanaged, no use). The second section—Database—gives an overview of the input of the six axes, which are listed in Figure 2.13.

Figure 2.13—The six axes of GLADIS: Four biophysical axes (light gray) and two socioeconomic axes (dark gray)

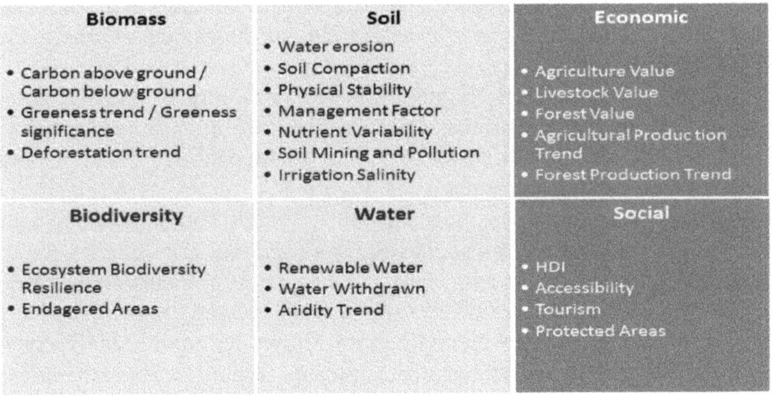

Biomass	Soil	Economic
• Carbon above ground / Carbon below ground • Greeness trend / Greeness significance • Deforestation trend	• Water erosion • Soil Compaction • Physical Stability • Management Factor • Nutrient Variability • Soil Mining and Pollution • Irrigation Salinity	• Agriculture Value • Livestock Value • Forest Value • Agricultural Production Trend • Forest Production Trend
Biodiversity	**Water**	**Social**
• Ecosystem Biodiversity Resilience • Endagered Areas	• Renewable Water • Water Withdrawn • Aridity Trend	• HDI • Accessibility • Tourism • Protected Areas

Source: Nachtergaele et al. 2010.

13 Link to GLADIS WebGIS: http://www.fao.org/nr/lada/index.php?option=com_content&view=article&id=180&Itemid=168&lang=en

14 Forest, agriculture, grassland, shrubs, rainfed crops, crops, wetlands, urban, sparse vegetated areas, bare areas, open water, no data.

The Analysis section shows ecosystem status, which corresponds to its actual degraded status, and ecosystem processes and trends for each axis. The status gives a summarized picture of the different input variables, whereas the trend combines two datasets from either two years or a time series (as used for the Greenness Trend, in which the NDVI time series of the GLADA approach [extended to 2006] was incorporated). In addition to information on land use, the degradation process and pixel- and country-level trends are shown. Regarding the land degradation process, a margin of 0–50 shows land degradation, whereas a margin of 50–100 shows the improvement of ecosystem conditions. Within the land degradation status maps, the status itself is described with a margin of 0 (worst) to 100 (best).[15] The final section shows the Land Degradation Index—or the simplified output—which is a selection of summarized land degradation statuses, processes, and impact indexes within six maps: the Ecosystem Service Status Index (ESSI), the Biophysical Status Index (BSI), the Land Degradation Index (LDI), Goods and Services severely affected, the Biophysical Degradation Index (BDI), and the Land Degradation Impact Index (LDII).

The ESSI (Figure 2.14) shows the actual state of goods and services provided by ecosystems, calculated by combining the four biophysical status axes (biomass, soil, water, and biodiversity) and the two socioeconomic status axes (economic and social status). Figure 2.14 shows that Sub-Saharan Africa and Australia both have a low status of ecosystem services. In these two regions, a reduction in ecosystem goods and services (due to land degradation, for example) will have higher negative pressure on population and land than, for example, in Scandinavia, the northwestern United States, and southern Canada, where the ESSI is high.

The BSI (Appendix B, Figure B.3) proceeds similarly to the ESSI, but factors only biophysical variables (status axes of biomass, soil, water, and biodiversity) into the output map, excluding socioeconomic ones. Goods and services severely affected (Appendix B, Figure B.4) represent the most severely affected areas, threatened by a huge decline of goods and services provided by an ecosystem. Therefore, a threshold rule was used that accounts for the critical values of the six axes.[16] This map (Figure 2.15) was superimposed on the overall processes of declining ecosystem services by considering the combined value of each process axis in the radar trend diagram, shown with the LDI (Nachtergaele et al. 2010). In addition to negative impacts, Figure 2.15 also shows improvement of land degradation, as seen in the Sahelian zone in Sub-Saharan

15 Although this feature seems to be interesting, it is not applicable to the national level, as determined by GLADIS itself, with a "warning to the users of GLADIS" occurring on the LADA website in October 2010. Therefore, unfortunately, this feature is useless.

16 Critical values of the six axes: Biomass < 25, Soil < 37.5 (This value coincides with a loss of 25 tons per hectare.), Water < 25, Biodiversity < 25, Economy < 25, Social/Cultural < 25.

Africa; this zone is faced by higher precipitation rates and an upcoming economic performance. The BDI (Figure 2.16), which only incorporates the biophysical axis, shows a more negative picture of land degradation as compared with the LDI. Due to the exclusion of socioeconomic aspects, we may conclude that an improvement of socioeconomic conditions is taking place in several countries, thus lowering land degradation. Yet this predication should be made carefully and with a close reference to every single input within the axis.

GLADIS also underlines the linkage between population pressure and land degradation, especially the strong relationship between poverty and land degradation. Using the database on global subnational infant mortality rates (IMR) and population density from CIESIN, we can get information on this linkage through the LDII (Figure 2.17).

Referring to Nachtergaele et al. (2010), 42 percent of the very poor live in degraded areas, as compared with 32 percent of the moderately poor and 15 percent of the nonpoor. Moreover, poverty was emphasized for being notably a rural problem, which is definitely not a new finding.

The GLADIS approach, with its WebGIS and the combination of biophysical as well as socioeconomic determinants, offers a broad but not exact view of global hot spots of land degradation. GLADIS emphasizes a close link between land degradation and poverty (Blaikie and Brookfield 1987; Barbier 1997, 2000; Duraiappah 1998) (see Box 2.2).[17]

GLADIS is not amenable to the global or national level, and the authors warn against using any information for policy strategies on the national or local level. The global scale generalizes the analysis; in addition, in some even-larger areas, such as Argentina,[18] it does not correctly represent the status of land degradation in a region. Results contradict other regional studies, such as the ones by Vlek, Le, and Tamene (2008, 2010). The GLADIS approach is limited due to the availability of global data with sufficient detail and resolution. For example, the analysis on "Trends of Agricultural and Forestry Production" (axis on economic determinants) contains a trend for national agricultural production, which takes into account livestock plus cropped agriculture for 1990 and 2003 by FAOSTAT (FAO Statistics), and a trend for forestry trends, which compares data of 1990 and 2006. "This gives rough estimates of trends in production over the 15-year

17 Infant mortality and child malnutrition both are proxies for poverty and welfare of an area (CIESIN 2010 http://sedac.ciesin.columbia.edu/povmap/methods_global.jsp). Within the GLADIS approach, a CIESIN global poverty dataset on infant mortality rates on a national level was given as an indicator for poverty. Data from the Demographic and Health Surveys, the Multiple Indicator Cluster Surveys, and the national Human Development Reports were incorporated in this dataset.

18 According to a presentation at the Technical LADA Meeting, September 6–14, 2010 in Wageningen, Netherlands; by the LADA Argentina Team. Available at www.fao.org/nr/lada/index.php?option=com_docman&task=doc_download&Itemid=165&gid=622&lang=en.

period" (Nachtergaele et al. 2010: 40) but does not illustrate an exact calculation, even if the earliest date corresponds.

Figure 2.14—Ecosystem Service Status Index, GLADIS

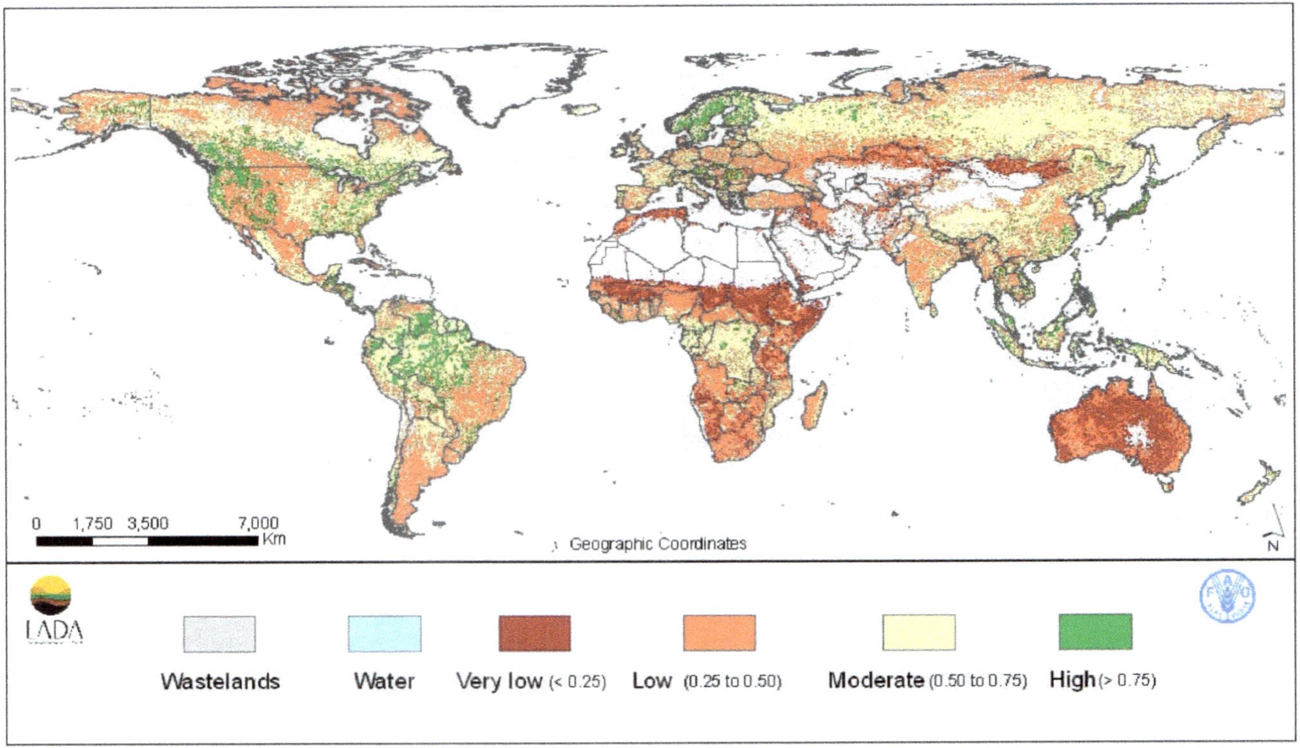

Source: Nachtergaele et al. 2010.

Figure 2.15—Land Degradation Index (LDI), GLADIS

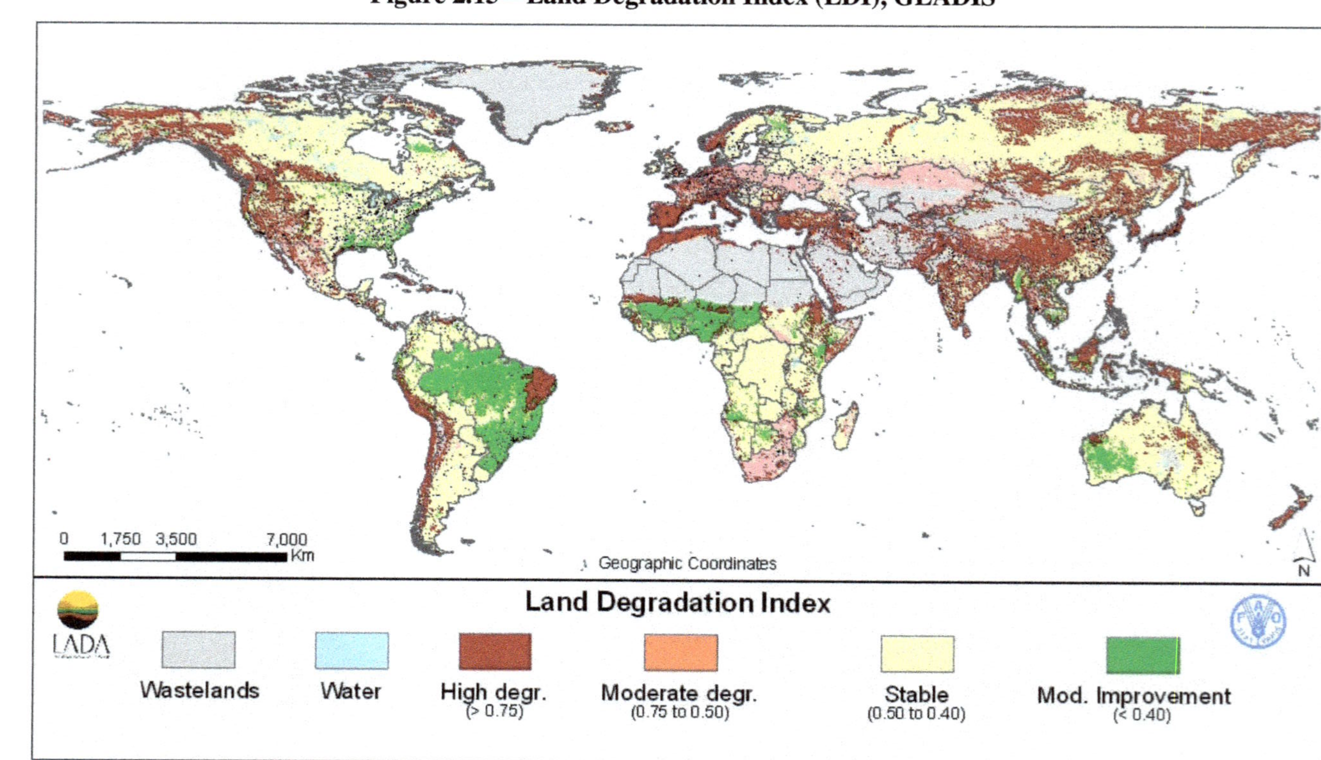

Source: Nachtergaele et al. 2010.

Figure 2.16—Biophysical Degradation Index (BDI), GLADIS

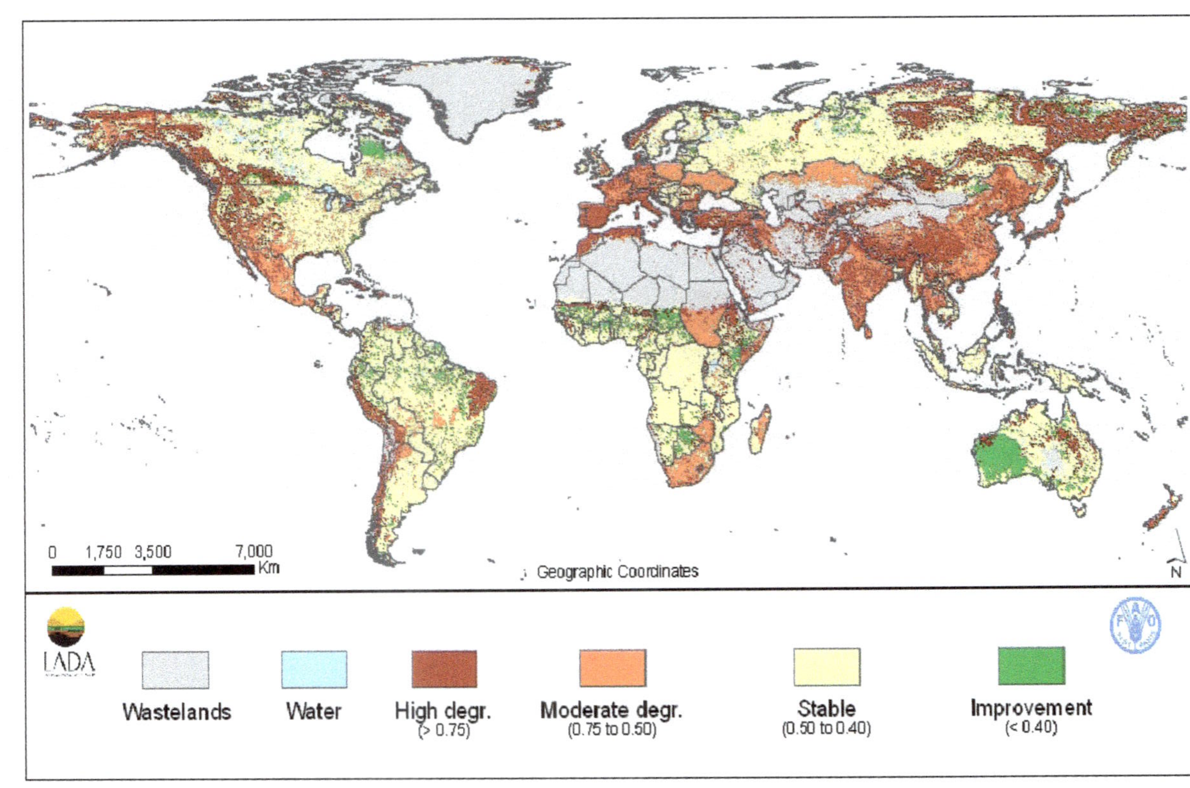

Source: Nachtergaele et al. 2010.

Figure 2.17—Land Degradation Impact Index (LDII), GLADIS

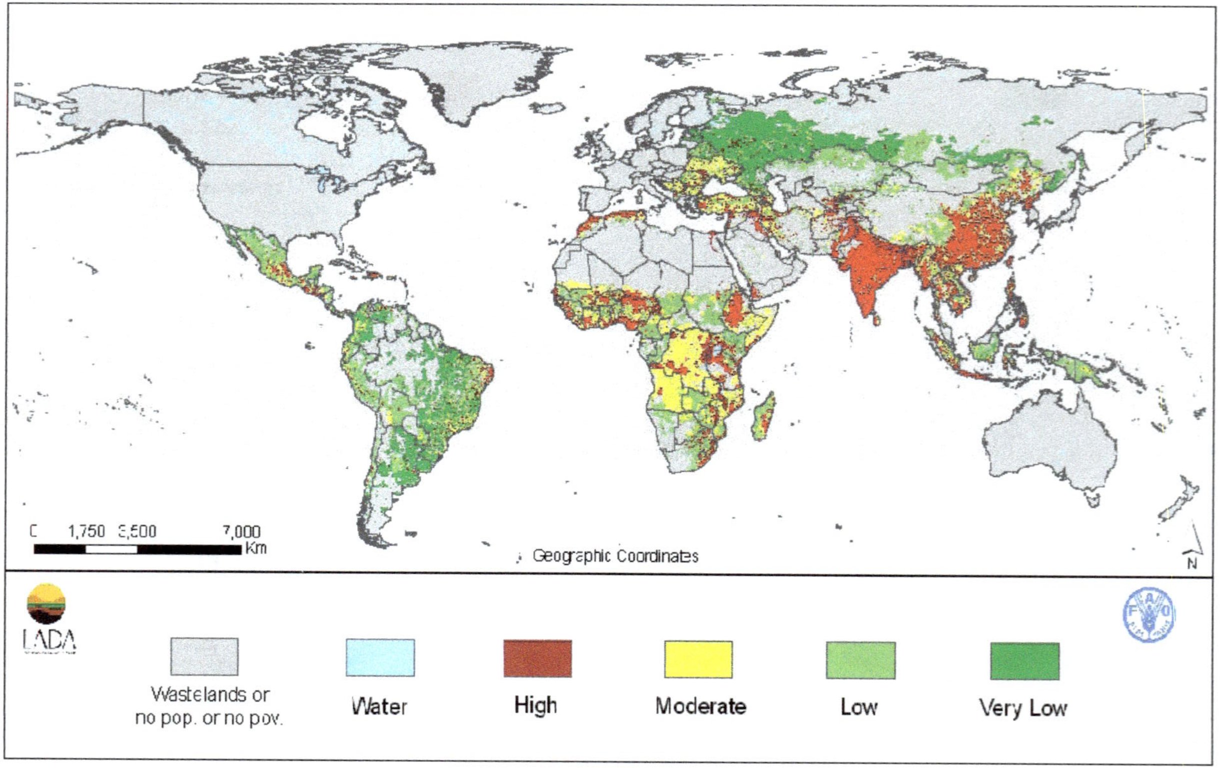

Source: Nachtergaele et al. 2010.

Box 2.2—Socioeconomic determinants of land degradation according to GLADIS

Within the GLADIS approach, six axes were used to get a broad and interdisciplinary view of land degradation. The Economic Indicators for Ecosystems consist of the Economic Production Status of Ecosystems, the Livestock Economic Status, the Forest Economic Status, and Trends of Agricultural and Forestry Production.

Within the Economic Production Status for agricultural land, two variables were used: an average crop yield for the year 2000—estimated in Global Agro-Ecological Zones (GAEZs)[19]—and a statistical database, mainly from FAOSTAT, on major agricultural crops. The average crop yield was measured in the international Geary-Khamis price[20] of 2000/01. Within the resource database, rainfed areas were distinguished from irrigated areas. The Livestock Economic Status includes information on climate regions (polar, desert, tropical, boreal, subtropical, and temperate), as well as on regional weighted averages of cattle and small ruminants shown in a livestock intensity map, according to Nachtergaele and Petri (2008).[21] Data on 2006 national forest GDP (Lebedys 2008) was the input for the global map on Forest Economic Status. A classification value from 0 (low) to 100 (high) for the year 2006, incorporating the country gross value, which was added to the forest sector, indicates the value of this index.Within the analysis maps, the axis inputs were calculated and combined to depict both land degradation *status* and land degradation *process*. Land degradation trends include trends on national agricultural production analyzed with livestock and crop data from 1990 and 2003, taken from FAOSTAT data. Moreover forestry trends were taken into account by comparing data on forestry production in U.S. dollars from 1990 and 2006 (Lebedys 2008). The Economic Production Status analysis for agricultural land reveals high economic output in countries such as the Netherlands, Belgium, the United Kingdom, France, and Germany relative to many developing countries, which are situated in drylands and which have low economic output. The land degradation trend analysis shows an increasing trend of crop and livestock production in Sub-Saharan Africa, Brazil, and several parts of Southeast Asia. A low trend is also clearly seen in Central and Eastern Europe. The overall trends of agricultural and forest production depict high values (an increasing trend) in Brazil, Sub-Saharan Africa, and the eastern parts of China, and low trends in Norway, Japan, and Indonesia. These results should be seen as being related to the input factors that concentrate on agricultural production, livestock, and forest areas. In general, drylands have a lower agricultural output and only limited forest areas; therefore, they of course show low values in the status analysis. The trend analysis takes into account different time series—agricultural production of 1990 and 2003, and forestry trends comparing a dataset of 1990 and

---

19 According to Fischer, Van Velthuizen, and Nachtergaele (2000:4), the GAEZs provide "a standardized framework for the characterization of climate, soil, and terrain conditions relevant to agricultural production." Therefore several determinants, such as the length of the growing period and the latitudinal thermal climates, were incorporated in this approach. Moreover, the GAEZs depict limitations in climate, including soil and terrain resources, which do have an impact on crops.

20 The Geary-Khamis price is an average price method that "entails valuing a matrix of quantities using a vector of international prices. The vector is obtained by averaging national prices across participating countries after they have been converted to a common currency with purchase power parities (PPP) and real final expenditures above the basic heading" (UN 1992: http://unstats.un.org/unsd/methods/icp/ipco_htm.htm).

21 The livestock intensity map is based on a map of Global Livestock Production Systems (Thornton et al. 2002) and data on cattle and small ruminants (Wint & Robinson 2007).

Box 2.2—Continued

2006. It should be questioned whether this analysis is still representative because of the different periods used, even if the starting points of the time series are the same.Global accessibility, tourism, protected areas, and the Human Development Index (HDI) are building the sixth axes on Social and Cultural Provisions of Ecosystems.

A map on global accessibility was prepared by Nelson (2008) for the World Bank's *World Development Report 2009*. With this map, the concentration of economic activity, as well as global accessibility, should be depicted to get a view of the benefits resulting from the concentration of production. Human displacement could be shown with infrastructure facilities.[22] Furthermore, the Environmental Systems Research Institute (ESRI)[23] software ArcGIS provides a cost–distance function that was used for this assessment. The reason for incorporating data on tourism, as in the Status of Accessibility of Ecosystems map, was that tourism is closely linked to environmental effects, such as the construction of general infrastructure and therewith the degradation of environmental resources. In addition, tourism awareness can be raised for ecosystems that could support the protection of natural areas and, in turn, increase their economic importance. Data on protected areas, based on the World Database on Protected Areas (WDPA 2008), "provide major benefits and social goods, including education, recreating tourism, conservation, and also the protection of vital services such as the provision of clean water" (Nachtergaele et al. 2010). A combined map—based on the tourism map and the map of social and cultural services by protected areas—was produced to depict the social and cultural services provided by ecosystems. Due to a lack of former datasets, trends of social and cultural services were difficult to depict with the given database on tourism and protected areas. Therefore, the HDI calculated by UNEP on the national level was chosen. This index is undoubtedly useful for representing social and cultural indicators for a country; however, a difficulty occurs if hot spots should be depicted with this dataset, because HDI data are only available on the national level.

Comparing two datasets, the trend analysis was calculated by subtracting the 2007 HDI rank number from the 1991 HDI rank number. Because countries can have slight increases and decreases within the HDI ranking over several years, the GLADIS approach builds groups in an effort to exclude these false alarms. A scale from 0 to 50 represents a downward trend, whereas a scale from 50 to 100 shows an upward trend.

The axis on social and cultural services is a good example of putting emphasis on the need for a uniform global assessment of land degradation. In former assessment methods, which neglect the incorporation of or a strong link to socioeconomic data with land degradation assessment, the selection of different data inputs is an advantage. But indeed, data should be comparable and should not switch between different times to get representative results.

The GLADIS approach of combining multiple indicators into maps lacks a more detailed description of why and how the selected indicators affect land degradation and how the generated maps can be interpreted. A combination of indexes simplifies the presentation of land degradation. However, the simplification hides differences, rendering some results less useful. For example, the

22 The ESRI software provides information on populated places, roads, railways, navigable rivers, water bodies, shipping lanes, land cover, urban areas, and elevation.

23 ESRI is the biggest software producer of geographical information systems.

use of the Human Development Index (HDI), an abstract indicator of a number of socioeconomic factors reported at national level, masks policy-relevant indicators that affect ecosystem services; thus, using results at the subnational level is limited. The results also do not say which HDI component needs to be changed to improve ecosystem services.

Mapping land degradation on a global level has definitely made advances. Table 2.3 gives a global land degradation assessment overview of some projects. From a general view focusing on climate conditions and depicting hot spots of desertification vulnerability by the calculation of the aridity index (AI),[24] the emphasis today is more on ecosystem goods and services in places where human life depends on and incorporates more than only biophysical determinants. It seems that the clearer the definition of land degradation, the more precise the assessment and, hence, the mapping of this process. GLADIS is the first mapping assessment that uses an interdisciplinary approach and that includes biophysical and socioeconomic determinants. Moreover, the definition of land degradation includes a time component that was not emphasized before. The crucial determinants of land degradation are slow variables that lessen the quality of an ecosystem's biophysical and socioeconomic determinants. The determination of *slow variables* is used by Reynolds et al. (2007) in the Dryland Development Paradigm (DPP) dealing with the interlinkage of biophysical and socioeconomic determinants in a coupled human-environment system. This discussion strengthens the need for a coherent and interdisciplinary assessment which should also be mentioned along with global mapping of land degradation.

When analyzing the current situation of land degradation, which includes the decrease in vegetation cover in particular, the focus is predominantly on visible indicators of consequences of the process (see the study reviews in Appendix A, Table A.1 and Table A.2). The satellite data do not include forms of degradation that cannot be detected remotely. Remote sensing is therefore limited to an evaluation of an aggregated outcome (vegetation cover) that is the result of various interacting factors on the ground, with one of those being land degradation. As observed earlier, although NDVI may indicate land degradation, it may also be misleading if factors other than land degradation lead to vegetation change. Recent approaches that take into account socioeconomic determinants of land degradation or improvement have attempted to address past weaknesses. For example, GLADIS combined multiple socioeconomic factors and biophysical measurements into indexes; however, depending on the factors chosen and their combination and weighting, the results changed. A more systematic approach and theoretical underpinnings are still needed to determine

---

24 Aridity index (AI) = Precipitation/Potential evapotranspiration.

which factors to select, how they interact and influence each other, and how they affect vegetation.

The general lack of data in developing countries makes land degradation assessment in a broad view sometimes difficult. However, availability of satellite imagery data has generally alleviated this data dearth problem in developing countries. Methods to assess land degradation are as manifold as the process itself. The use of radar and microwave remote sensing must be integrated more often in actual land degradation assessment techniques. A global approach is needed that uses standardized methods and a bottom-up technique that starts at the local level, enabling the adaptation of global analysis data to the local level. Global monitoring is still a challenge. As pointed out earlier, there is still a lack of precise data on the global level. Global maps on land degradation and desertification do give good overviews, but, as pointed out within the GLASOD, GLADA, and GLADIS approaches, information cannot be transferred to the local level. This local-level information is needed for policymakers and for more adapted research on land use management.

Table 2.3 summarizes the strengths and weaknesses of the land degradation and improvement approaches used in the past and reviewed in this study.

Table 2.3—Global land degradation assessment studies

Project and Duration	**What is monitored?**	**Techniques used and strength(s)**	**Extent/severity of land degradation**	**Scale/ resolution of maps**	**Limitations**	**End product**
UNCOD (1977)	"Estimated" desertification; desertification hazard	Expert opinion: Limited number of consultants with experience in drylands	35%, or 3,970 million hectares, of Earth's surface is affected by desertification	Data not georeferenced.	Subjective, due to expert opinion; no georeferenced data	Desertification hazard map
GLASOD (1987–1990)	Human-induced soil degradation; status of soil degradation, including the type, extent, degree, rate, and causes of degradation within physiographic units	Expert opinion (more than 250 individual experts): Data were later digitized to a GIS-database—four types (water erosion, wind erosion, chemical LD, physical LD) and four degrees of LD (light, moderate, severe, very severe). *Global assessment taken into account, not only drylands*	65% of the world's land resources are degraded to some extent; 1,016–1,036 million hectares of drylands are experiencing LD.	Produced at a scale of 1:10 million; 1:5 million FAO soil map was also integrated in the study (data for 1980–1990)	Subjective ,due to expert opinion; focus on soil degradation, does not include all types of LD; maps are too rough for national policy purposes.	One map showing four main types of LD (water erosion, wind erosion, chemical degradation, physical degradation) and four degradation severities (light, moderate, strong, extreme)
ASSOD (1995)	Regional study of GLASOD (Assessment of Soil Degradation in South and Southeast Asia); data from 17 countries	Expert opinion (national institutions): Analysis due to the use of SOTER; data stored in database and GIS.	> 350 million hectares of ASSOD area, or 52% of the total susceptible dryland area	1:5 million (data for 1970–1995)	Lack of available data; difficult to distinguish between human- and natural-induced degradation; subjective, due to expert opinion	Variety of thematic maps with degree and extent of land degradation

Table 2.3—Continued

Project and duration	What is monitored?	Techniques used and strength(s)	Extent/severity of land degradation	Scale/ resolution of maps	Limitations	End product
SOVEUR (Soil Vulnerability Assessment in Central and Eastern Europe) (1998)	Regional Study of GLASOD (data from 13 countries)	Providing a database based on SOTER and the use of expert opinion, as in GLASOD; based on quantitative satellite data rather than expert opinion	About 186 million hectares, or 33%, of the area covered by the SOVEUR project, is degraded to some extent.	1:2.5 million (data for 1973–1998)	Link to environmental and social pressures is missing.	Provision of an environmental information system with a SOTER database for the 13 countries under consideration
UNEP (World Atlas of Desertification) *(WAD used the GLASOD output.)*	1st edition (1992): depiction of land degradation in drylands; 2nd edition (1997): assessment of several indicators: vegetation, soil, climate, and so on, plus combating measurements and socioeconomic variables, such as poverty and population data; 3rd edition: 2011/2012	Based on the GLASOD approach, which used expert opinion	*—See GLASOD.—*	Using GLASOD data with 1:10 million resolution	Focus on drylands; subjective, due to expert opinion	World Atlas of Desertification, including maps on soil erosion by wind and water, chemical deterioration; case studies focus on Africa and Asia (due to ASSOD)
WOCAT (since 1992)	Soil and water conservation (SWC); conservation approaches and technologies to combat desertification should be mapped; network of SLM specialists	Expert opinion: Case studies in 23 countries on six continents with three questionnaires on mapping, technologies, and approaches; more objective, due to the use of SOTER; SWC technologies; cost of SWC data can be used to assess cost of preventing or mitigating land degradation.	Focus is put on SWC to guide investments to those areas where they are most needed and most effective (points show SWC method).	Small-scale world map (1:60 million), for showing current achievement of SWC	Good national case studies that cannot be extrapolated to global level. Mapping still in development; first draft exists.	Detailed maps at (sub)country level; first draft of global overview of achievements in preventing and combating desertification exists (in collaboration with FAO and by request of the Biodivesity Indicators Partnership, Convention on Biological Diversity (COP10)

Table 2.3—Continued

Project and duration	What is monitored?	Techniques used and strength(s)	Extent/severity of land degradation	Scale/ resolution of maps	Limitations	End product
USDA-NRCS (1998–2000)	Desertification vulnerability; vulnerability to wind and water erosion and "human-induced" wind and water erosion; analysis of soil moisture and temperature regimes, population density, serious conflicts with risk to desertification	GIS/modeling with FAO soil map, climate database; population data from CIESIN; depicting land quality classes with given datasets	34% of the land area is subject to desertification; 44% of the global population is affected by desertification.	1:100 million; *minimum scale* 1:5 million FAO soil map: 1:5 million	Socioeconomic data takes into account only population densities—life is only classified as "human-induced." *Positive*: Categorizing land quality classes; seems as if NRCS distinguished between desertification and LD.	Several maps on global soil climate map, land quality, desertification vulnerability, and human-induced desertification vulnerability; water and wind erosion and human-induced water and wind erosion
GLADA (2000–2008)	Soil degradation, vegetation degradation, national assessment (LADA), global assessment of degradation and improvement (GLADA); over a certain period (1981–2003, extended to 2006)	Remote sensing (GIMMS dataset of 8-km-resolution NDVI data); input of SOTER in support of general NDVI methodology. Based on quantitative satellite data; not on expert opinion; correlation of land degradation with socioeconomic data	24% of the land area was degraded between 1981 and 2003 (80% of the degraded area occurred in humid areas).	Grid cells of 32 km^2 Data for 1981–2003 (extended to 2006) LADA: 1:500,000–1:1 million	Primarily monitoring of land cover; analyzing trends—lack of information on the present state; degradation before 1981 and in areas where visible indicators could not be monitored yet were not included	Identifying hot spots of degrading and improving areas
MA (2005)	Global ecosystem services	14 studies (global, regional, and subregional); based on remote sensing and other data sources, with georeferenced results compiled into a map with grid cells of 10 x 10 km^2	60% of global ecosystem services degraded or used unsustainably	Grid cells of 100 km^2 Data within the 1980–2000 period	Different studies used for MA with different definitions of LD and different time periods of assessment; no economic assessment of ecosystems	Relationship between ecosystems and human welfare

Table 2.3—Continued

Project and duration	What is monitored?	Techniques used and strength(s)	Extent/severity of land degradation	Scale/ resolution of maps	Limitations	End product
GLADIS (2010)	Mapping of the status of LD and pressures applied to ecosystem goods and services by using six axes of biophysical and socioeconomic determinants (biomass, soil, water, biodiversity, economics, and social)	Remote sensing, GIS, (LADA database); Modeling; "Spider diagram approach;" integration of more than population data in socioeconomic determinants; broad analysis of the process of land degradation	Relationship: LD and poverty: 42% of the very poor live in degraded land, 32% of the moderately poor, and 15% of the nonpoor.	5 arc minute (corresponds to 9 km x 9 km)	Combining national and subnational data, taking into account different periods of the different inputs; lumping of many indicators loses focus and attribution	Global Land Degradation Information System; Provision of general data and analysis on LD due to a WebGIS

Sources: Oldeman, Hakkeling, and Sombroek 1991b; Oldeman 1998; Thomas and Middleton 1994; Van Lynden, Liniger, and Schwilch 2002; MA 2005; Bai et al. 2008b; Nachtergaele et al. 2010.

Notes: LD = land degradation; SLM = sustainable land management; GIMMS = Global Inventory Modeling and Mapping Studies; CBD = Convention on Biological Diversity; BIP = Biodiversity Indicators Partnership.

2.2 Types of Land Degradation

Land degradation can be classified according to different types: physical, chemical, and biological processes. These types do not necessarily occur individually; spiral feedbacks between processes are often present (Katyal and Vlek 2000). Physical land degradation processes refer to erosion; soil organic carbon loss; changes in the soil's physical structure, such as compaction or crusting and waterlogging (that is, water accumulates close to or above the soil surface). Chemical processes, on the other hand, include leaching, salinization, acidification, nutrient imbalances, and fertility depletion. According to Hein (2007), soil erosion, whether induced by water or wind, involves translocation of topsoil from one place to another and represents the most important land degradation problem. Pimentel (2006) estimated that about 30 percent of the global arable land has been severely eroded in the past 40 years. Soil productivity is lost through reduced rooting depth, removed plant nutrients, and physical loss of topsoil. One important feature of soil erosion by water is the selective removal of the finer and more fertile fraction of the soil (Stocking and Murnaghan 2005).

Figure 2.18 gives an overview of methods used to assess different types of land degradation. Although most of the assessments are done on the local level, some methods are available for assessing types of land degradation on large areas, such as the analysis of vegetation or the monitoring of water turbidity modeling. Usually erosion is a natural soil-forming process that can be accelerated by human actions (Katyal and Vlek 2000). Determining factors of soil erosion are rainfall (erodibility), vegetation (cover), topography, soil properties (erodibility), slope inclination, and exposure (sun, shadow), as well as socioeconomic factors like population density and severity of poverty (de Graaff 1993).

Soil compaction, another form of physical land degradation, is common in areas using heavy machinery or areas with high livestock density. Waterlogging and salinization are mainly caused through inefficient irrigation systems, where improperly lined canals lead to seepage and result in a rise of water tables. The GLASOD study estimated that salinity accounted for about 4 percent of the degraded land area (see Table 2.1).

Figure 2.18—Methods for the assessment of land degradation

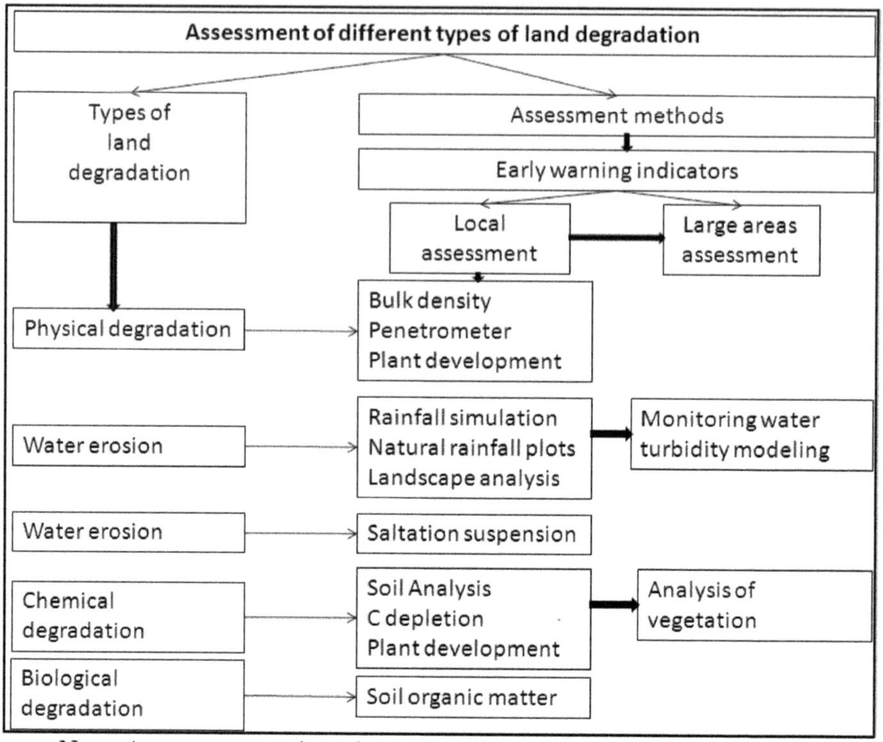

Assessment of different types of land degradation

Types of land degradation

Assessment methods

Early warning indicators

Local assessment

Large areas assessment

Physical degradation → Bulk density; Penetrometer; Plant development

Water erosion → Rainfall simulation; Natural rainfall plots; Landscape analysis → Monitoring water turbidity modeling

Water erosion → Saltation suspension

Chemical degradation → Soil Analysis; C depletion; Plant development → Analysis of vegetation

Biological degradation → Soil organic matter

Note: A penetrometer is an instrument that measures the hardness of a substance.

Source: Modified from Castro Filho et al. 2001.

The usual depth of salts in soils cannot be maintained, and the resulting salinity in topsoils leads to decreased plant growth if it is not diluted or washed away by rainfall (Katyal and Vlek 2000). It is estimated that about 20 percent of irrigated area is affected by salinity (Pitman and Lauchli 2004).

Soil nutrient mining is also an important problem in countries that apply limited amounts of fertilizer. Tan, Lal, and Wiebe (2005) estimated that about 56 percent of area planted with wheat, barley, rice, and maize experienced soil nutrient mining, which led to a yield reduction of 27 percent in 2000. Developing countries account for about 80 percent of the global soil nutrient mining. As shown in Table 2.4, in the 1990s, South America and Africa, respectively, accounted for about 50 percent and 34 percent of areas with some form of soil nutrient depletion. However, soil nutrient depletion in Sub-Saharan

Africa has been more severe than any in other region due to the limited use of fertilizer (Henao and Baanante 1999).

Table 2.4—Extent and severity of global soil nutrient depletion, 1990–1999 (in million hectares)

Location	Light	Moderate	Strong	Total	% of total
Africa	20.4	18.8	6.2	45.4	34
Asia	4.6	9.0	1.0	14.6	11
South America	24.6	34.1	12.7	71.4	53
Other regions	2.8	1.2	0	3.9	3
Globe	52.4	63.1	19.9	135.3	

Source: Tan, Lal, and Wiebe 2005.

Biological processes include rangeland degradation, deforestation, and loss in biodiversity, including loss of soil organic matter (which also affects the physical and chemical properties of the soil) or of flora and fauna populations or species in the soil (such as earthworms, termites, and microorganisms) (Scherr 1999). Vegetation degradation is a long-term loss of natural vegetation, with a decrease in biomass and ground cover of perennial native vegetation (Coxhead and Oygard 2007; Katyal and Vlek 2000). Hence, changes in the structure and botanical composition of plants constitute vegetation degradation, which can also occur naturally due to sparse native vegetation. However, in contrast to induced vegetation degradation, natural degradation is typically gradual and often reversible (Katyal and Vlek 2000). Major causes of the destruction of natural vegetation are fires, the use of heavy machines, fuelwood extraction, and overgrazing by livestock stands (Katyal and Vlek 2000). The destruction of natural vegetation directly leads to reduced residues from plants, leading to organic matter loss (FAO 1994; Stocking and Murnaghan 2005). Vegetation degradation is studied using the NDVI. As reported earlier, Africa south of the equator accounted for the largest loss of vegetation between 1981 and 2003 (Bai et al. 2008b).

2.3 Causes of Land Degradation

Proximate Causes

As discussed in the conceptual framework of action and inaction, proximate causes of land degradation are those that directly cause land degradation. These are further divided into biophysical factors and unsustainable land management practices.

The biophysical proximate causes of land degradation include topography, land cover, climate, soil erodibility, pests, and diseases. Soil erosion is a function of slope length, land cover, and steepness (Wischemeier 1976; Voortman, Sonneveld and Keyzer 2000). Steep, long slopes are vulnerable to severe water-induced soil erosion if they have poor land cover with no physical barriers to prevent erosion. The severity of water- and wind-induced soil erosion is higher if land clearing is done on mountain slopes. Pests and diseases, such as invasive species, lead to loss of biodiversity, loss of crop and livestock productivity, and other forms of land degradation.

Climatic Conditions

Climate directly affects terrestrial ecosystems. For example, dry, hot areas are prone to naturally occurring wildfires, which, in turn, lead to soil erosion, loss of biodiversity, carbon emission, and other forms of land degradation. Strong rainstorms lead to flooding and erosion, especially if such rainstorms occur during the dry season in areas with poor land cover. Rainfall patterns such as low and infrequent rainfall and erratic and erosive rainfall (monsoon areas) lead to a low soil-moisture content, which then leads to reduced plant productivity and high runoffs, resulting in erosion and salinization because salts in the soil surface are not leached into deeper soil layers (Safriel and Zafar 2005). Furthermore, drought-prone areas are more likely to be naturally degraded (Barrow 1991). A consequence of elevated levels of carbon dioxide, caused by global warming, increases desertification processes (Ma and Ju 2007). Other consequences of climate change include reduced rainfalls, which lead to changes in land use or to a reduction in land cover due to prolonged droughts (see Box 2.3 for more details). In some cases, climatic impacts are of sufficient intensity to induce ecological land degradation, or degradation that naturally occurs without human interference. However, anthropogenic activities often trigger or exacerbate such ecological land degradation (Barrow 1991).

Box 2.3—The relationship between climate change and land degradation

Climate change and land degradation are related through the interactions of land surface and the atmosphere. These interactions involve multiple processes, with impact flows running in both directions—from the land surface to the atmosphere and vice versa. These complex processes take place and vary simultaneously (WMO 2005). The feedback effects between climate change and land degradation are not yet fully understood.

Climate change affects land degradation because of its longer-term trends and because of its impacts on the occurrence of extreme events and increased climate variability. Climate change trends include the increase in temperature and a change in rainfall patterns, which are two determinants in the creation and evolution of soils, most notably through their impact on vegetation distribution. Climate variability holds the potential for the most severe human impacts. For instance, the occurrence and severity of droughts has been related to actual declines in economic activity, whereas gradual increases in mean temperature have not. In Sub-Saharan Africa, in particular, climate variability will affect growing periods and yields and is expected to intensify land degradation and affect the ability of land management practices to maintain land and water resources (Pender et al. 2009) in the future. However, it must also be noted that climate change is not solely a negative influence on land degradation—for instance, agroclimatic conditions are expected to improve in some areas. Simultaneously, land degradation affects climate change through (1) the direct effects of degradation processes on land surface, which then affects atmospheric circulation patterns, and (2) the effects of land degradation on land use, with land use changes then affecting the climate. In these complex interrelationships between climate change and land degradation, sustainable land and water management (SLWM) can play a crucial mitigating role. Notably, research has already shown the links between soil carbon sequestration and its impacts on climate change and food security (Lal 2004). Soil carbon sequestration transfers atmospheric carbon dioxide in the soils, hence mitigating its climate change impacts. Increasing soil carbon stocks, in turn, has a positive impact on crop productivity, at least past a certain minimum threshold (World Bank 2010, 77). Thus, SLWM practices that sequester large amounts of soil carbon can provide a win–win–win solution in the issues of climate change, land degradation, and some of their human dimensions, such as food security. Examples of such practices include no-till farming, cover crops, manuring, and agroforestry (Lal 2004). The extent of these win–win–win situations and the conditions under which they can be realized are areas that require more systematic research. Just as climate change and variability will affect different regions in different ways, so too will their consequences relating to LD vary in general and to specific types of land degradation in particular. Further, the linkages between land and climate systems hold important keys to the valuation of the costs of LD and of land conservation or restoration.

Topography

Steep slopes lead to land degradation. Fragile, easily damaged soils located along steep slopes are often associated with soil erosion if vegetation cover is poor. Lands located in drylands—as well as lowlands close to the sea, exposed coastal zones, or areas prone to extreme weather and geological events (such as volcanic activity, hurricanes, storms, and so on)—show low resilience and are thus vulnerable to erosion, salinization, and other degradation processes (Safriel and Zafar 2005).

Unsustainable Land Management

Land clearing, overgrazing, cultivation on steep slopes, bush burning, pollution of land and water sources, and soil nutrient mining are among the major forms of unsustainable land management practices.

Underlying Causes

Policies, institutions, and other socioeconomic factors affect the proximate causes of land degradation. We discuss key underlying causes of land degradation, some of which were discussed earlier. A brief discussion will be given for such factors that have already been discussed.

National Level Policies

As discussed, policies have a large impact on land management practices. Policies could have a direct or indirect impact on land users' behavior. For example, current efforts by the Costa Rican government to promote and invest carbon sequestration have set an exemplary success story in developing countries. Since 1997, Costa Rica started investing significantly in payment for ecosystem services as part of its forest and biodiversity policies. Such a policy has made Costa Rica a payment for ecosystem services (PES) pioneer in developing countries (Pagiola 2008). Likewise, more than one-third (with actual adoption rate in parentheses) of crop area in Argentina (58 percent), Paraguay (54 percent), Uruguay (47 percent), and Brazil (38 percent) is under conservation agriculture (Kassam et al. 2009). Farm subsidies have also contributed to higher adoption of fertilizer in several developing countries, including India and several African countries (Heffer and Prud'homme 2009). In 2001, farm subsidies in Organization for Economic Cooperation and Development (OECD) countries were about $235 billion (FAO 2003; Anderson 2009), a level that has contributed to an adverse impact on international trade against developing countries (Anderson 2009).[25] Subsidies have also contributed to environmental pollution arising from low fertilizer prices, which lead to overapplication (Mulvaney, Khan, and Ellsworth 2009). Fertilizer subsidies have also led to overuse of input or other farmer behavior, which have been harmful to the environment. Such farmer behavior is heavily influenced by national-level policies.

---

25 The United States alone used $35 billion on farm subsidies in 2007 (Edwards 2010), whereas western Europe and Japan contributed the largest share of farm subsidies (Anderson 2009).

Local Institutions

As discussed earlier, local institutions are important drivers of land management practices. Strong local institutions with a capacity for land management are likely to enact bylaws and other regulations that could enhance sustainable land management practices (FAO 2011). As pointed out earlier, national-level policies —such as decentralization—and the presence of internal and external institutions to build the capacity of the local institutions on land management play key roles. In general, top-down policies are found to lead to alienation and land degradation.

International Policies and Strategies

International policies through the United Nations and other organizations have influenced policy formulation and land management in all countries of the world. In the past 40 years, international policies and initiatives have increasingly been oriented toward sustainable development (Sanwal 2004) and have been affecting country-level and community-level land management practices. Among the most remarkable international sustainable development initiatives are the Rio Summit of 1992, the Millennium Summit of 2000, the 2002 Johannesburg Summit on Sustainable Development, and global research synthesis efforts, such as the Millennium Ecosystem Assessment or the reports of the Intergovernmental Panel on Climate Change. An international initiative that directly addresses land degradation is the UNCCD (Box 2.4). Environmental policies at both the national and international level are increasingly becoming common across countries. Nearly every country has an Environmental Protection Authority responsible for regulating and enforcing environmental laws and regulations. Globally, there are 500 multilateral environmental agreements (MEAs), which have been ratified by a majority of the countries in the world (UNEP 2011). Realization of the important urgency of protecting the environment has also grown across all countries, due to the increasing pollution and environmental degradation in general and the global awareness and promotion of sustainable development (Sanwal 2004).

Box 2.4—Successes and challenges of UNCCD

The UNCCD has been ratified in 194 countries. The membership underlines the worldwide popularity of the convention. The design of the national action plans (NAPs) revolves around a participatory bottom-up approach that seeks to empower local communities in implementing the NAPs. In addition, the NAPs are supposed to learn from research and to create synergies with existing programs. The NAPs also emphasize the need for accountability, which is a reflection of the desire to show their effectiveness in combating desertification. Among the key successes that have influenced land management in developing countries are the following:

1. The promotion of global cooperation to address land degradation and desertification
2. A greater awareness of LD and the need to take action to address the problem
3. Endorsement of a participatory approach and emphasis on a decentralized implementation of actions.

This approach has helped developing countries that ratified the UNCCD to decentralize land management and to use local knowledge—an aspect that plays a key role in sustainable land management. The UNCCD has set a prime example of the community-based implementation of UN conventions (Bruyninckx 2004) and other conventions that have attempted to follow the bottom-up approach in designing national-level initiatives. For example, the United Nations Framework Convention on Climate Change (UNFCCC) national adaptation program of action (NAPA) also followed the bottom-up approach (Bruyninckx 2004).

Despite these successes, UNCCD faces the following major challenges:

1. The actual implementation of the NAPs has been minimal, largely due to the limited capacity of developing countries.

2. There is a limited commitment from governments to commit resources to implement the activities proposed. The NAPs have largely been donor funded, which has placed them under project mode so that they have not been integrated into other national policies and programs. As a result, in most countries, program implementation and planning has been tuned to respond to perceived donor expectation, rather than reflecting the country's policies and priorities. Only a few countries have been able to form long-term action plans and to mainstream them with national programs and policies.

3. There has been limited mainstreaming of the NAPs in other international and national programs. For example, the Convention on Biological Diversity (CBD), NAPA, and NAPs are largely implemented by one ministry in many countries. Efforts to mainstream NAPs have been made through the national steering committees (NSCs), but these have remained weak with limited clout over other ministries.

4. Unlike its sister conventions (NAPA and NBSAP), NAP has largely remained a developing countries program, with little implementation in developed countries or middle-income countries.

5. The actions proposed in many NAPs fail to address the fundamental role that institutions and policies play in land management. In cases where institutions such as strengthening of local governments are addressed, the resource allocation has been absent or limited. In addition, the NAPs do not generally try to seek actions to change policies and institutions.

International policies and strategies have played a key role in fostering sustainable development in developing countries. Multilateral and bilateral donor support to natural resource management accounts for a large share of expenditure in land and water resources in Sub-Saharan Africa (Anonymous 2006).For example, donors accounted for 70 percent of the total expenditure in SLM in Mali and Uganda (World Bank 2008b, 2011). International support of natural resource management in other developing regions is also significant. In 2008, about $6 billion was given by Overseas Development Administration (ODA) countries to developing countries, with Asia accounting for 54 percent of the support and Sub-Saharan Africa accounting for 21 percent of support (OECD 2010). Such support has significantly influenced land management in developing countries.

Such support has also been directed to enhancing productivity, environmental protection, and a score of other natural resource management issues. For example, the Asian Green Revolution was an international initiative aimed at increasing agricultural productivity to meet the increasing demand for food. The Green Revolution contributed to a reduction of the conversion of land use to agricultural production (Hazell 2010). Using 1950 cereal yield as a benchmark, Borlaug (2000) estimated that without the Green Revolution and other yield-enhancing technologies, 1.8 billion hectares of land of the same quality would have been required—instead of the 660 million hectares that was used—to produce the harvest of 2000. This underscores the large influence of international policies and strategies on land management.

Access to Markets

Land users in areas with good market access are likely to receive higher produce prices and to buy agricultural inputs at lower prices, which could create an incentive to invest in land management (Pender, Place and Ehui, 2006). Nkonya, et al. (2008a) observed more positive nitrogen and phosphorus balances for households closer to roads than for those farther from roads. However, high market access also creates alternative livelihood opportunities and increases the opportunity cost of labor, which could lead to a lower propensity to adopt labor-intensive land management practices, such as the use of manure and soil and water conservation practices (Scherr and Hazell 1994). Warren (2002) introduced the notion that land degradation is contextual, pointing out that the fields worked by households with many options were found to be more affected by erosion than those of households with fewer options. Similarly, other authors have indicated that through increased off-farm employment opportunities, the cost for labor increases, leading to less application of labor-intensive conservation measures. Households may also determine an optimal rate of land degradation, at which depleting land and investing in off-farm employment or education may pay off in the future—rather than preventing or mitigating land degradation. Similarly, manure might not be applied to crop plots if farmers have alternative uses for that manure, such as heating and cooking. This all shows that the key economic driving forces affecting land management decisions and the trade-offs between economic and ecological goals at the farm household level need to be better understood.

Alternative livelihoods could also allow farmers to rest their lands or to use nonfarm income to invest land improvement. Nkonya et al. (2008a) found that nonfarm employment was associated with a greater propensity for fallow, soil nutrient balance, and lower soil erosion. This finding suggests that other factors affecting land management must be taken into account when considering the impact of access to markets.

Access to Agricultural Extension Service

Access to agricultural extension services enhances the adoption of land management practices. Clay et al. (1996) found that extension services are strongly and significantly associated with less erosive forms of land use in Rwanda; a similar finding was made by Paudel and Thapa (2004) in Nepal.[26] Depending on the capacity and orientation of the extension providers, access to extension services could also lead to land-degrading practices. For example, studies in Uganda and Nigeria have shown that farmers with access to agricultural extension services were more likely to use improved seeds and fertilizer but less likely to use organic soil fertility management practices (Benin et al 2007; Nkonya et al 2010). Extension services, public research, and trainings do not always require the introduction of new technical solutions; rather, sometimes the encouraged use of simplified techniques, such as vegetation barriers and stone buds (Kassie et al. 2008), can have large impacts. Unfortunately, agricultural extension services remain limited in developing countries.

Population Density

Empirical evidence has shown a positive relationship between population density and land improvement (see, for example, Bai et al. 2008a; Tiffen, Mortimore, and Gichuki 1994), which supports Boserup's (1965) agricultural intensification under high population density. Contrary to this, empirical evidence has also shown a positive relationship between land degradation and population density (see, for example, Grepperud 1996). Conditioning factors such as agricultural marketing influence the impact of population density on land management. The impacts of some of these have been discussed above.

Between 1700 and 1999, the per capita cropland availability worldwide has fallen from 0.39 hectare to 0.22 hectare (Katyal and Vlek 2000). A growing population of raising per capita income has an increasing need for food. Limited land resources often lead to a division into smaller pieces of land when land is

26 See Vanclay (2004) for more details on the social principles in agricultural extension.

divided in the inheritance process.[27] Figure 2.19 illustrates the declining trends in per capita arable land in four regions of the world. Rapidly declining per capita land area is associated with the conversion of forest land and other land use into cropland. Expansion into more fragile land has also been a common challenge accompanying declining per capita arable land area.

Figure 2.19—Trend of per capita arable land across regions

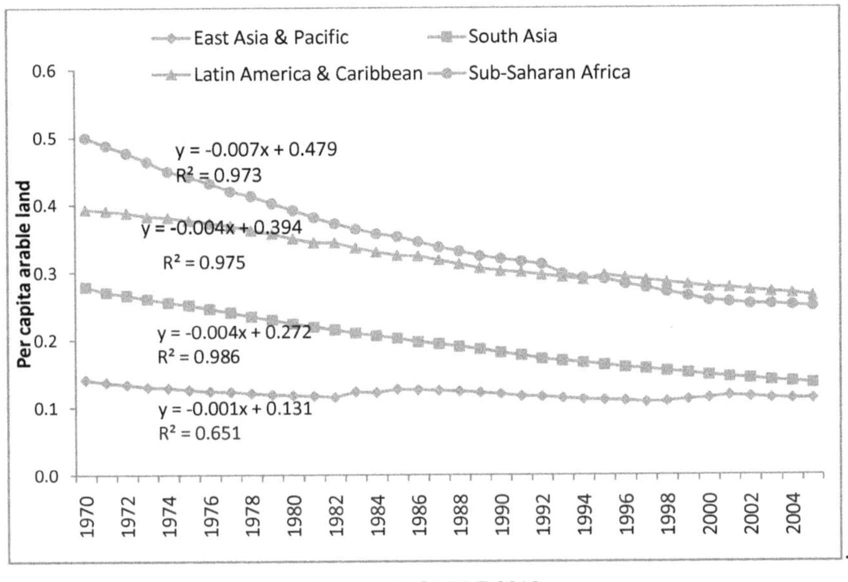

Source: FAOSTAT 2010.

Increasing pressures on agricultural land have resulted in much higher nutrient outflows and the subsequent breakdown of many traditional soil-fertility maintenance strategies, such as bush-fallow cultivation and the opening of new lands. Fallow periods have decreased, and soil regeneration through long-term fallow can no longer be maintained (Giller et al. 1997; Padwick 1983).

Migration, either as outmigration or immigration, also has an impact on degradation. Outmigration of men into urban centers leaves women and old people in rural areas. Women may face difficulties in accessing agricultural

27 Muchena et al. (2005) cited a number of researchers who have not found population growth to contribute to widespread soil degradation and fertility decline, such as studies by Mazzucato and Niemeijer (2001) in Burkina Faso; by Tiffen, Mortimore, and Gichuki (1994) in Kenya; and by Walker and Ryan (1990) in the semiarid areas of India. According to these studies, villages have been managing natural resources without giving rise to irreversible degradation, despite population pressure.

inputs, and usually the most able-bodied workers leave the rural areas. Some authors argue that the migrant populations moving in contribute to continued degradation, because due to land shortages, they have to cultivate marginal areas and adopt inappropriate farming technologies in their struggle for a living (Gachimbi et al. 2002; Ndiritu 1992). A study in Bhutan showed that the arrival of refugees first led to deforestation but that host communities later formed community-based forest management groups, which then led to better forest management than before the refugee influx (Birendra and Nagata 2006). Likewise, Whitaker (1999) observed that the arrival of refugees in Tanzania from Rwanda and Burundi provided cheap labor and higher demand for agricultural products, which all led to better land management. Bai et al. (2008b) has shown a positive correlation between population density and NDVI. The results further show the impact of other conditioning factors.

Land Tenure

Secure land tenure and land rights—or at least long-term user rights—are vital for providing an incentive to invest in soil and water conservation measures. Diverse systems of landownership, tenure, and land rights exist across continents, with different degrees of tenure security. Insecure land tenure can lead to the adoption of unsustainable land management practices. Using panel survey data collected from farming households, Kabubo-Mariara (2007) showed that for Kenya, property right regimes and population density affect both the decision to conserve land and the type of conservation practices used by farmers. The results further suggest a positive correlation between land tenure security and population density. Regarding the application of manure, evidence from farmers using own and borrowed land for cultivation show that manure application is more frequently applied on the former than on the latter (Gavian and Fafchamps 1996), which underlines the importance of long-term incentives for single users, who are not necessarily dependant on the system as a whole. Although there have been several empirical investigations into the relationship between land tenure and investment, existing evidence is largely inconclusive (see Brasselle et al. 2002). A growing body of literature has demonstrated the failure of many land titling efforts, especially in Africa, to improve land management, increase agricultural productivity, or reduce poverty and conflict (see, for example, Atwood 1990; Migot-Adholla, Hazell, Blarel, and Place 1991; Place and Hazell 1993; Platteau 1996; Deininger 2003). Empirical evidence has shown that farmers holding land under insecure land rights may plant trees or do other investments to enhance their security (Besley 1995; Place and Otsuka 2002; Brasselle et al. 2002). In addition, evidence from Sub-Saharan Africa has shown that land investment of farmers holding land under customary land tenure was either comparable to or greater than land investment of farmers holding land under more secure leasehold or freehold land tenure (Toulmin and Quan 2000; Deininger 2003; Nkonya et al. 2008a).

Land-titling programs have also affected nomadic livelihoods, which are hardly compatible with the landownership concept. Sedentarization of nomads, as well as other agricultural policies allocating crop production in rangelands, put pressure on the already-shrinking grazing lands, the quality of declining rangelands, and carrying capacity (Thomas and Middleton 1994). Sedentarization inhibits nomads from flexibly adjusting according to their transhumance routes on less-degraded areas. The forced concentration of farmers and livestock leads to increased conflicts due to competition for the natural resources, thus shifting agricultural use to marginalized lands and consequently to overgrazing.

Deininger et al. (2003) argued that the impact of tenure insecurity varies across investments, having encouraged planting trees but discouraged investments in terraces. They further showed that the mere perception of more stable property rights did encourage construction of terraces, indicating that people's decisions are strongly affected by perception.

The process of assigning property rights to individual users might also leave out other legitimate users; therefore, careful analysis of bundle of rights is needed (Meinzen-Dick and Mwangi 2008; Schlager and Ostrom 1992). This is particularly important as property rights on land should be thought of as a web of interests, with many different parties having a right to withdrawal, management, exclusion, and alienation, leading to different types of right holders. The formalization of property rights has historically led to a cutting of this web, creating more exclusive forms of rights over the resource (Meinzen-Dick and Mwangi 2008). Such "cutting of the web" is not only politically sensitive but also relatively costly if registration, cartography, and so forth are done with rigor. In addition, introducing formalized private titles would not remove disincentives to invest on borrowed land (Gavian and Fafchamps 1996).

Despite the empirical evidence supporting preservation of the customary land tenure systems, however, application of the traditional land rights system becomes more and more difficult, and pressures develop to ensure formal ownership of existing cultivated land (FAO 2001). The customary land tenure systems in many traditional communities are biased against women; in many societies, women are not allowed to inherit or possess land, though they are responsible for the agricultural household production (Syers et al. 1993). Women often do not have access to production inputs like fertilizers, and soil nutrient depletion is the consequent result, leading to land degradation (Bossio, Geheb, and Critchley 2010). Degradation is then an outcome of policy and institutional failures, which include a lack of well-defined, secure, tradable property rights (Ahmad 2000).

To summarize, the point is not that land titling or other approaches to formalizing land rights are never useful. Indeed, favorable impacts of land titles on land management have been found in many countries, including Thailand (Feder 1987), Brazil (Alston et al. 1995), Honduras (Lopez 1997); and Nicaragua (Deininger and Chamorro 2004). Rather, the point is that the impacts of land tenure interventions, such as land titling, are highly context dependent. In addition, efforts to promote improved land tenure systems must be well-suited to the context in which they are applied if they are to help improve land management, reduce poverty, and achieve other objectives.

Infrastructure Development

Transport and earthmoving techniques, like trucks and tractors, as well as new processing and storage technologies, could lead to increased production and foster land degradation if not properly planned (Geist and Lambin 2004). MNP/OECD (2007) estimated that infrastructure development and land use change are the major factors contributing to biodiversity loss (measured using mean species abundance). Motor pumps and boreholes, or the construction of hydrotechnical installations, such as dams and reservoirs, often lead to high water losses due to poor infrastructure maintenance and high leakage rates. Consequently, the water cycle is affected irreversibly.

Poverty

The debate over the impact of poverty on land degradation remains inconclusive. One school of thought posits the vicious cycle of poverty–land degradation, which states that poverty leads to land degradation and that land degradation leads to poverty (see, for example, Way 2006; Cleaver and Schreiber 1994; Scherr 2000). In what Reardon and Vosti (1995) termed *investment poverty*, poor land users lack the capital required to invest in land improvement. Neither labor nor capital resources are available to invest in land conservation measures, such as green manuring or soil conservation structures (FAO 1994). Because farmers cannot afford inputs such as fertilizer, pesticide, or irrigation equipment, the productivity of the land declines. The low productivity puts pressure on marginal lands, which are cultivated to add to the family income. Poor farmers tend to be associated with marginal lands and low yields (Rockström, Barron, and Fox 2003), which is manifested in their lack of financial means, poor health status, and outmigration by men. Safriel and Adeel (2005) describe this process as a downward spiral of low productivity and land degradation in which "poverty is not only a result of desertification but a cause of it" (Safriel and Adeel 2005: 646).

Another school of thought maintains that the poor, who heavily depend on the land, have a strong incentive to invest their limited capital into preventing or mitigating land degradation if market conditions allow them to allocate their resources efficiently (de Janvry, Fafchamps, and Sadoulet 1991). Subsistence farmers may deplete their soils less rapidly due to limited outflow of soil nutrients off the farm. For example, Nkonya et al. (2008a) observed a negative relationship between livestock endowment and nitrogen balance. Similarly, they observed a negative relationship between soil erosion and livestock endowment.

The preceding discussion shows the complex relationship of the proximate and underlying causes, which makes it hard to generalize using a simple relationship of one underlying factor with a proximate cause of land degradation. The results imply that one underlying factor is not, in itself, sufficient to address land degradation. Rather, a number of underlying factors need to be taken into account when designing policies to prevent or mitigate land degradation.

2.4 Associations between Potential Drivers of Land Degradation

Land cover change is the most direct and pervasive anthropogenic factor used to determine land improvement or degradation (Vitousek 1994; Morawitz et al. 2006). Several studies have used the normalized difference vegetation index (NDVI) and other measures based on the NDVI, as an indicator of changes in ecosystem productivity and land degradation. In chapter 2, we have described and analyzed the limitations and criticisms related to the use of this measure. However, NDVI remains the only dataset available at the global level and the only dataset that reliably provides information about the condition of the aboveground biomass. Mindful of all the limitations, we follow an approach similar to Bai et al. (2008b) and investigate the relationship between changes in NDVI (from 1981 to 2006) and some key biophysical and socioeconomic variables (Table 2.5).

Data and Methods

The analysis of NDVI change is based on data derived from the Global Inventory Modeling and Mapping Studies (GIMMS), which supply NDVI data from July 1981 to December 2006. In the GIMMS dataset, two NDVI observations are available for each month and therefore each year is composed of 24 sets of global data[28] (Tucker, Pinzon, and Brown 2004; Pinzon, Brown, and Tucker 2005; Tucker et al. 2005). For each pixel (which is the unit of observation in our analysis) and year, the average NDVI is computed then averaged across two time periods: (1) the baseline of 1982 to 1986 and (2) the end line of 2002 to 2006. Subtracting pixel by pixel the baseline from the end line NDVI value we obtain the change in average NDVI.

28 This includes one maximum composite value from the first 15 days of the month and one from day 16 to the end of the month (Tucker, Pinzon, and Brown 2004).

NDVI values in agricultural areas are strongly dependent on farmers' production decisions (for example, crop choices, fertilizer usage, and irrigation). As a consequence, the relationship between NDVI and land degradation for the observations that cover agricultural areas is tenuous. We decided therefore to eliminate from the dataset those observations that cover areas where agriculture is the predominant land use. In order to perform this operation, we used data from the Spatial Analysis Model (SPAM). This model is used to identify at the global level areas where agriculture is predominant. Specifically, we identified all the locations (roughly one pixel of ten-by-ten kilometers at the equator) where cropland represents 70 percent or more of the land use. The NDVI observations that fall in these areas were dropped from the study.

The choice of the biophysical and socioeconomic variables used to explain the change is strongly dictated by data availability. Unfortunately, important information on poverty, cost of access, road networks, and urban areas is not available as panel data and therefore could not be included in the analysis. The variables used (Table 2.5) include precipitation, population density, government effectiveness, agricultural intensification (proxied by fertilizer application), and country gross domestic product (GDP).[29] To avoid influence by abnormal years, we take an average of four consecutive years for the baseline and end line periods. However, not all data were available for the baseline and end line periods. In such cases, we used time periods closest to the two NDVI time periods (Table 2.5).

We know a priori changes in precipitation have a strong effect on NDVI and expect a positive correlation between positive changes in precipitation and NDVI. The impact of population density on land degradation is ambiguous. While the induced innovation theory (Hayami and Ruttan 1970; Boserup 1965) predicts that farmers will intensify their land investment as population increases, other studies have suggested more land degradation in areas with greater population density (Cleaver and Schreiber 1994; Scherr 2000). As will be seen in chapter 5, however, institutions influence the impact of population density and other drivers on land degradation.

The effects of agricultural intensification on NDVI and land degradation are not clear a priori. Intensification could slow down the conversion of forest land into agricultural land (corresponding to an observed increase in NDVI). Fertilizer application increases soil carbon (Vlek et al. 2004), which could correspond to an increase in NDVI. Hence, the relationship between NDVI and agricultural intensification is ambiguous.

On the other hand, we expect to see a strong positive correlation between government effectiveness—or a government's capacity to implement policies

---

29 As discussed below, we also included the squared value of GDP to account for possible non-linearity.

with independence from political pressures and with respect to the rule of law (Kaufmann, Kraay, and Mastruzzi 2009)—and NDVI. Past studies have shown that government effectiveness, which also reflects the quality of civil services and a government's commitment to implementing its policies, is strongly correlated with the level of democracy experienced by a country (Kurzman et al. 2002; Adsera et al. 2003; Barro 1999), which is strongly related to better natural resource management. Highly centralized governments are less effective at local levels and tend to concentrate decisionmaking of natural resource management in the central governments, which leads to greater resource degradation (Anderson and Olstrom 2008).

Table 2.5— Selected variables used to analyze relationships with NDVI

Variable	Resolution	Baseline	End line	Source of data
NDVI	8km x 8km	1982–84	2003–06	Global Land Cover Facility (www.landcover.org), Tucker, Pinzon, and Brown 2004); NOAA AVHRR NDVI data from GIMMS
Precipitation	0.54° x 054°	1981–84	2003–06	Climate Research Unit (CRU), University of East Anglia www.cru.uea.ac.uk/cru/data/precip/
Population density	0.5° x 0.5°	1990	2005	CIESIN (2010)
Government effectiveness	Country	1996–98	2007–09	Worldwide Governance Indicators: http://info.worldbank.org/governance/wgi/index.asp
Agricultural Intensification	Country	1990–92	2007–09	FAOSTAT
GDP	Country	1981–84	2003–06	IMF: www.imf.org/external/pubs/ft/weo/2010/02/

Sources: Compiled by authors.

The relationship between economic growth, natural resource use, and sustainability has been extensively studied (Dinda 2004; Lopez and Mitra 2000), but the debate on the effects of economic growth on the health of the environment is ongoing. The most well-known attempt to capture the essence of this relationship is the environmental Kuznets curve (EKC). The EKC has an inverted-U shape quadratic curve (Grossman and Krueger 1991). The EKC model hypothesizes that environmental degradation first increases as the economy grows but later reaches a plateau and then decreases. (For empirical evidence in support of the EKC and a review of its opponents, see Dinda 2004.)

We included GDP in our explanatory variables to represent economic growth. Like Bai et al. (2008b), we first use a simple correlation analysis to assess the relationship between NDVI and the selected variables. Since such relationships could differ across regions, we disaggregate our analysis across the major regions defined by the United Nations. We also use an ordinary least-square regression analysis to establish the correlation of NDVI with all other variables simultaneously.

Correlation Analysis

Consistent with past studies (for example, Grepperud 1996), Table 2.6 shows a negative correlation between change in population density and NDVI in all regions except Sub-Saharan Africa (SSA), the European Union (EU), and Near East and North Africa (NENA).

Table 2.6—Correlation of NDVI with selected biophysical and socio-economic factors

Variable	East Asia	EU	Latin American Countries	Near East and North Africa	North America	Oceania	South Asia	SSA
Δ Population density	-0.03*	0.002	-0.01*	0.04*	-0.01*	-0.01*	-0.02*	0.01*
Δ Precipitation	-0.02*	-0.04*	0.17*	0.23*	-0.01*	0.09*	0.24*	0.13*
Δ Agricultural intensification	0.06*	-0.01*	0.20*	0.01*	-0.14*	-0.10*	0.14*	-0.01*
Δ GDP	0.06*	0.17*	0.03*	0.28*	0.14*	0.10*	0.21*	0.09*
Δ Government effectiveness	0.09*	-0.04*	0.24*	0.23*	-0.14*	0.10*	0.08*	0.10*

Source: Author's calculations.

Notes: Statistical significance codes: *significant at the 5% level. Δ = Change from end line to baseline period.

This is contrary to Bai et al. (2008), who observed a positive correlation between NDVI and population density on a global scale. Consistent with Bai et al. (2008b), however, population density was positively correlated with NDVI in the SSA, EU, and NENA regions. In SSA, population density is highest in the most fertile areas, such as mountain slopes (Voortman, Sonneveld, and Keyzer

2000). This leads to the positive correlation between NDVI and population density even in areas south of the equator, which have seen severe land degradation (Bai et al. 2008b). Figure 2.20 also shows that there was a positive correlation between population density and NDVI in central Africa, India, North America, and Europe. We also see an increase in NDVI accompanied with negative population density in Russia (Figure 2.20). Our results show that in all regions, GDP changes are positively correlated with NDVI changes (Table 2.6). Figure 2.21 also shows an increase of both GDP and NDVI in North America, Russia, India, central Africa (north of the equator), and China. This suggests the role ecosystems could play in economic growth.

Consistent with expectations, the correlation analysis showed that in most regions, government effectiveness is positively correlated with NDVI. It was negative only in the EU and North America, which is largely due to a decrease in government effectiveness during the period under review accompanied by an increase in NDVI in both regions (Figure.22). With the exception of the EU, North America, Oceania, and SSA, the correlation between agricultural intensification (proxied by fertilizer application) and NDVI is positive as expected (Table 2.6). The EU, North America, and Oceania have seen a decrease in fertilizer application (Figure 2.23), which could explain the apparent negative correlation with NDVI. In SSA, land conversion to agriculture is responsible for the declining NDVI.

Figure 2.20—Relationship between change in NDVI and population density

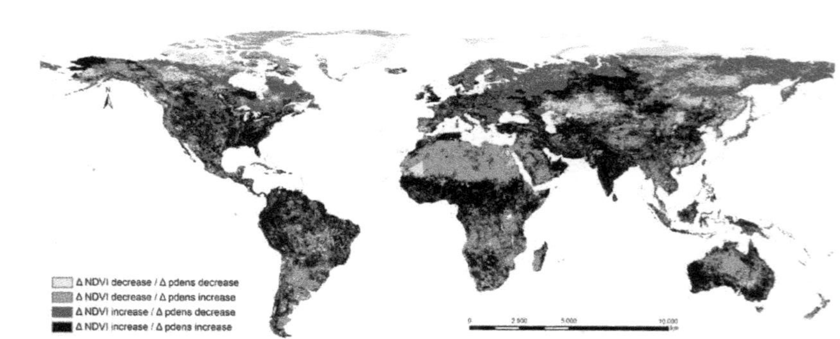

Figure 2.21—Relationship between GDP and NDVI

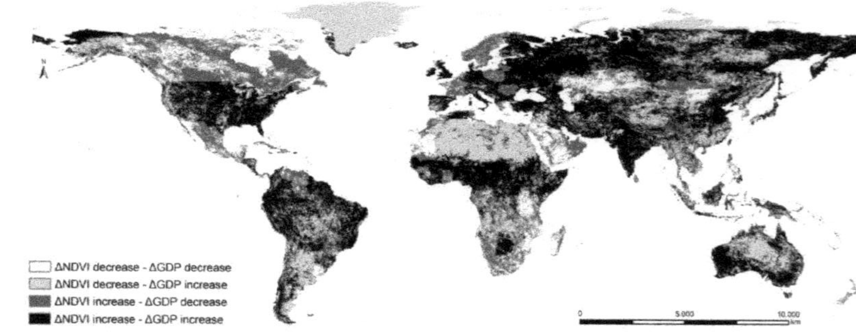

Figure 2.22—Relationship between government effectiveness and NDVI

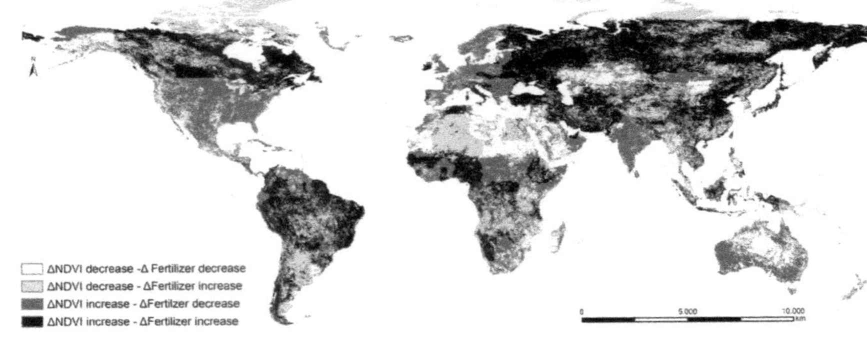

Figure 2.23—Relationship between fertilizer application and NDVI

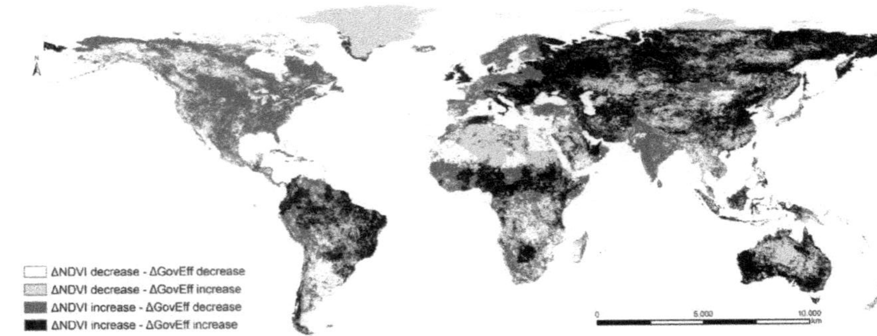

Regression Analysis

Table 2.7 reports the results of the regression at both the global level and disaggregated across the major regions to evaluate the robustness of the results. Oceania and North America are missing from the results due to the small number of countries in the regions, which leads to low variability of the national level explanatory variables and consequent problems with the estimation process. Due to the complex and multidirectional relationship between NDVI and the selected variables, the results are to be interpreted with extreme caution. Note that the results might not indicate a causal relationship but only an association of NDVI with the selected biophysical and socioeconomic variables. The effects of changes in population density are clearly less consistent across geographical areas. Generally, increases in population density correspond to decreases in NDVI values. This is not true for East Asia and Near East and North Africa. As expected, increments in precipitation are related to increases in NDVI. This is a result consistent across all geographical regions. The results for the effects of changes in agricultural intensification appear to be very robust. With the exception of the Near East and North Africa, agricultural intensification is associated with higher values of NDVI. The results for government effectiveness are also consistent and follow our expectations. With the exception of East Asia, greater government effectiveness is associated with higher NDVI values. The relationship between GDP growth and NDVI is less clear. Contrary to the environmental Kuznet curve, the GDP–NDVI relationship is U-shaped, suggesting that land degradation first declines as GDP increases and then increases beyond a threshold. This type of result is observed in only three regions: the European Union, Latin America, and the Near East and North Africa. In East Asia, South Asia, and Sub-Saharan Africa, however, the relationship between GDP and NDVI follows the expected environmental Kuznet curve.

Table 2.7—OLS regression of mean NDVI on selected biophysical and socio-economic variables

Variable	Global	East Asia	European Union	Latin American Countries	Near East and North Africa	South Asia	Sub Saharan Africa
Δ Population density	1.414×10^{-3}	2.042×10^{-3} *	-3.160×10^{-2} *	-6.575×10^{-3}	9.970×10^{-3} *	-2.381×10^{-3} **	-5.286×10^{-3} ***
Δ Precipitation	1.934×10^{-2} *	1.801×10^{-4}	7.763×10^{-3} **	2.002×10^{-2} *	6.025×10^{-2} *	2.873×10^{-2} *	1.514×10^{-2} ***
Δ Agricultural intensification	5.995×10^{-3} *	4.286×10^{-2}	8.102×10^{-2} ***	4.895×10^{-1}*	-8.370×10^{-3} *	6.893 *	7.312×10^{-3}
Δ GDP	-2.099×10^{-3} *	5.077×10^{-2} ***	-1.600×10^{-2}**	-4.185×10^{-1} *	-8.679×10^{-2} *	4.788 *	2.701***
Δ GDP^2	4.511×10^{-7} *	-1.627×10^{-5}**	9.53×10^{-6}	3.668×10^{-4} *	9.851×10^{-4} *	-6.825×10^{-3}	-8.865×10^{-2} ***
Δ Government effectiveness	4.880*	-1.937×10^{-2}	1.441×10^{-1}*	2.031×10^{-1} *	8.112*	-	7.975***
Constant	1.629×10 *	-1.817×10**	4.341×10 *	2.146×10*	6.487*	6.239*	5.669***

Source: Author's calculations.

Note: * significant to the 0.1% level; **significant at the 5% level; *** significant at the 1% level.

2.5 Effects of Land Degradation

On-Site Effects

The direct on-farm impacts of soil degradation on agricultural production can be experienced by farmers through declining yields, which are a result of the changes in soil properties (Clark 1996). A global study based on 179 experiments from around the world regarding the relationship between soil erosion and crop yield (den Biggelaar et al. 2003) provided the best available evidence of the impacts of soil erosion and crop productivity. As shown in Figures 2.24–2.26, yield losses of the three reference crops (maize, wheat, and millet) due to soil erosion are substantial, ranging from about 0.2 percent for millet to almost 0 for all three crops. Loss of yield productivity was larger in developing countries than in Europe or North America, thus showing the severe impact of land degradation on crop productivity and its variability across regions, which is partly due to mitigating practices taken to address degradation. For example, U.S. farmers apply fertilizer worth about $20 billion annually to offset soil nutrient loss due to soil erosion (Troeh et al. 1991).

Figure 2.24—Impact of soil erosion on wheat yield

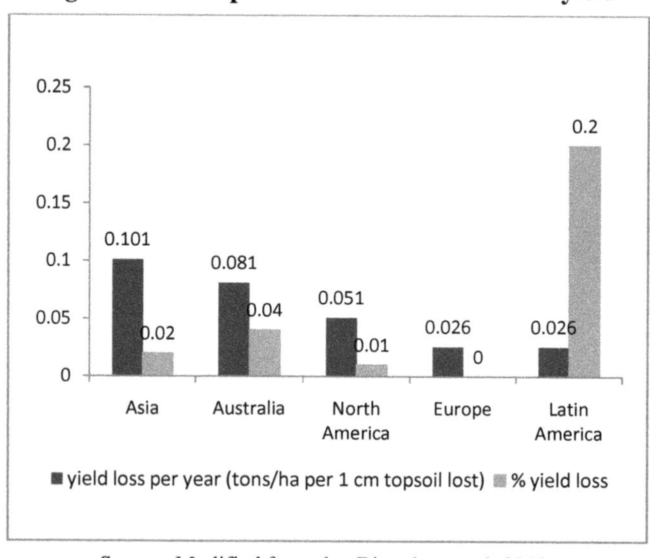

Source: Modified from den Biggelaar et al. 2003.

Figure 2.25—Impact of soil erosion on maize yield

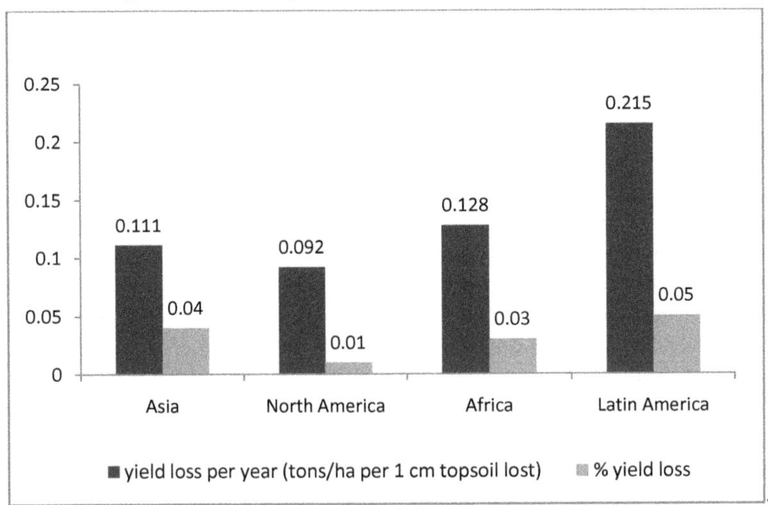

Source: Modified from den Biggelaar et al. 2003.

Figure 2.26—Impact of soil erosion on millet yield

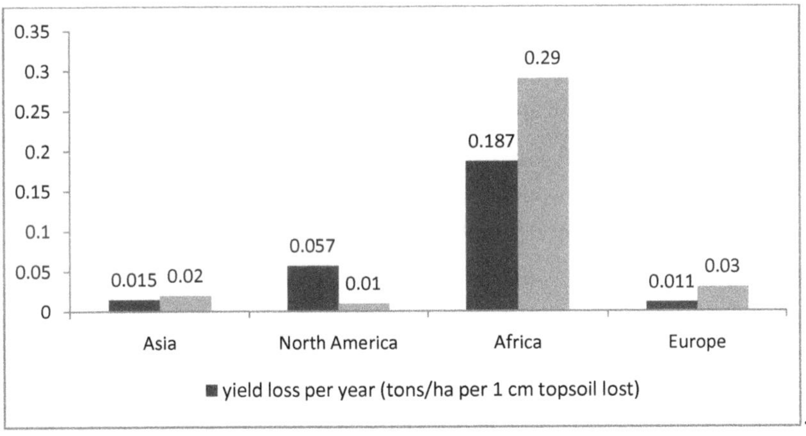

Source: Modified from den Biggelaar et al. 2003.

All theses mechanisms are closely interlinked and have spiral feedbacks on crop yields. However, it is important to mention that the extent to which erosion actually reduces yields depends on the types of crops, implying that the crop management system can have an influence on crop yields and the effects of land degradation. In the long run and in instances of serious degradation, the effects of land degradation might lead to temporary or permanent abandonment of plots and to a conversion of land to lower-value uses (Scherr and Yadav 1996).

According to Bossio, Geheb, and Critchley (2010), there is a strong link between land and water productivity, implying reduced water productivity due to land degradation, which leads to greater demand for agricultural water. Water quality and storage may both be reduced due to land degradation.

Socioeconomic on-site effects include the increase of production costs due to the need for more inputs to address the negative physical impacts of land degradation. Income losses arise as a consequence of erosion and land degradation, as farmers are not able to pay for inputs and to invest in improved land management methods (Bojö 1996). These costs can be measured as productivity losses through fertility and nutrient loss, soil loss through erosion, or a reduction in the vegetation cover—or even as changes in groundwater supply, loss of wood production, loss of grazing and hunting possibilities, carbon sequestration, nature conservation, and tourism.

Other ecosystem services are lost due to land degradation. For example, tree cutting reduces the availability of fuelwood, which in turn increases the labor input required for collecting fuelwood (Cooke et al. 2008). Degraded lands lead to loss of biodiversity, which in turn leads to reduction in other ecosystem services used by households. Soil erosion reduces the absorptive and storage capacity of water, which in turn increases the demand for water on eroded plots. Moderately eroded soils absorb 7–44 percent less water per hectare per year from rainfall than do uneroded soils (Murphee and McGregor 1991). An increase in the demand for irrigation water implies higher production costs, low yields and plant biomass, and consequently lower overall species diversity within the farm ecosystem (Walsh and Rowe 2001).

Overall, food security in particular is a major concern for households. Reduced land productivity leads to food insecurity.

Off-Site Effects

Land degradation may also have important off-site costs and benefits, including the deposition of large amounts of eroded soil in streams, lakes, and other ecosystems through soil sediments that are transported in the surface water from eroded agricultural land into lake and river systems. The deposits raise the waterways and make them more susceptible to overflowing and flooding; they also contaminate the water with soil particles containing fertilizer and chemicals. The beneficial off-site effects of soil erosion include the deposition of alluvial soils in the valley plains, which forms fertile soils and higher land productivity. For example, the alluvial soils in the Nile, Ganges, and Mississippi river deltas are results of long-term upstream soil erosion, and they all serve as breadbaskets in riparian countries (Pimentel 2006). The provision of fertile sediment on floodplains may decrease crop yield upstream while increasing yields in the alluvial valley plains (Pimentel 2006; Clark 1996).

The siltation of rivers and dams reduces reservoir water storage, leading to decreased water availabilities for irrigation and for urban, industry, and hydroelectricity uses. It also damages equipment and reduces flood control structures. Finally, it disrupts the stream ecology, decreases navigability of waterways and harbors, increases maintenance costs of dams, and shortens the lifetime of reservoirs. It is estimated that about 0.5 percent of annual water storage is lost annually due to sedimentation from soil erosion (White 2010). At regional levels, reservoir storage losses are shown in Figure 2.27. Central Asia experiences the largest annual loss of about 1 percent of storage capacity due to siltation.

Figure 2.27—Annual loss of reservoir storage capacity due to sedimentation

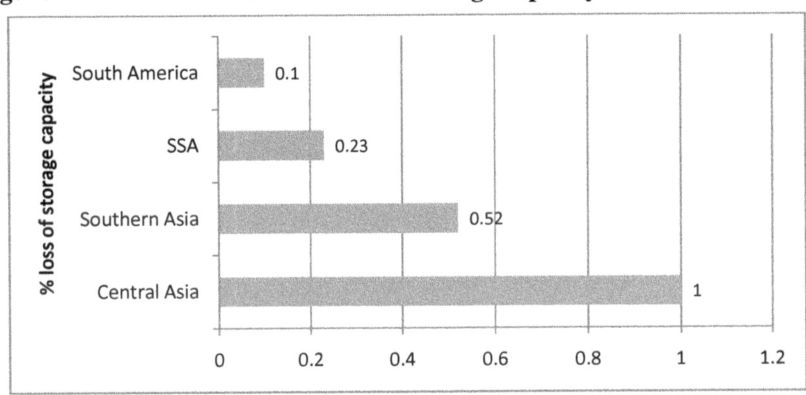

Source: White 2010.
Note: SSA = Sub-Saharan Africa

If soil is wind eroded, it causes health problems, as soil particles that are propelled by strong winds are abrasive and air pollutants. This can have an effect on health worldwide, as dust is transported over long distances, leading to higher costs of healthcare (Montanarella 2007). Furthermore, water quality can be influenced directly as concentrations of agrochemicals, metals, and salts are increased. When biomass carbon in the soil is oxidized due to soil erosion and a loss of biodiversity and biological activity, carbon dioxide is released into the atmosphere, contributing to global warming. This action can be seen as a feed-back mechanism, as global warming intensifies rainfall, which in turn increases erosion (Pimentel 2006).

As a consequence of increased land degradation, other natural resources, such as lime for neutralizing acidity or water for flushing irrigation salinity, are more in demand in order to repair the land. This leads to off-site pollution and further losses of productivity and amenity values (Gretton and Salma 1997).

Water pollution due to fertilizer use is also high. It is estimated that about 40 million tons of nitrogen and 10 million tons of phosphorus are deposited into water bodies annually (Corcoran et al. 2010; Rockström et al. 2009). Nutrient runoff causes eutrophication in lakes and pollutes coral reefs, leading to severe impacts on fish and human populations.

Further off-site effects refer to environmental services enjoyment of wild flora and fauna and other human activities, such as recreation and the amenity value of water resources. As biodiversity decreases, land becomes less resistant to droughts and requires more time to recover its productivity (Pimentel 2006). The loss of keystone species might affect the survival of other species, as well as the biological cycle within the ecosystem.

As the population grows, more food will be needed and more will be produced on marginal lands with low productivity, which has an effect on food security as well as on farm income and poverty rate (Eswaran, Lal, and Reich 2001). A decrease in productivity caused by land degradation also indirectly affects food security. As land degradation decreases the natural productivity of the soil, it has the potential to decrease production, or at least to increase production costs. These two effects, in turn, raise food prices and increase food insecurity and poverty. Poverty, as seen earlier, is itself a cause for land degradation. Hence the linkages between land degradation and poverty through the impacts on food and input prices have the potential to create vicious circles, with potential self-accelerating speed. However, higher food prices also offer the potential for improved adoption of conservation measures in agriculture by increasing their profitability (Pender 2009). Therefore, the complex interactions between land degradation and prices must be thoroughly examined on a case-by-case basis.

2.6 Summary

In this report, *land degradation* is defined as the change in productivity and the provision of ecosystem services, as well as in the human benefits derived from them. The terms land degradation and desertification are sometimes used interchangeably in the literature; however, the latter is strictly defined as land degradation in drylands. In this report, we prefer to use the term land degradation, which is more inclusive, as we refer to a global assessment that should cover all climate zones.

To assess the extent of land and soil degradation and desertification, a number of studies have been carried out, starting in the 1970s with the map of the status of desertification in arid lands by UNEP (1977), followed by maps on the desertification of arid lands (1983) and on global desertification dimensions and costs (1992), both by Texas Tech University. Going beyond drylands, Oldemann, Hakkeling, and Sombroek (1991a) undertook the GLASOD study, which indicated the extent and severity of soil degradation in all climates. The study was useful for raising attention to the extent and severity of soil degradation and contributed to the formulation of a number of global conventions and international and national land management development programs. It was also subject to criticism, because it was based on subjective expert judgments.

Those early studies largely focused on determining the biophysical forms of land degradation. As a response to GLASOD, WOCAT was initiated in 1992 and is still ongoing in order to document, monitor, and evaluate soil and water conservation measures worldwide. The LADA/GLADA (1981–2003) studies made use of GIS and remote sensing data to map land degradation. Recognizing the need to link (physical) land degradation to its underlying causes and its impact on humans, LADA/GLADA included socioeconomic variables such as poverty and population density in their maps. They found a positive relationship between land degradation and poverty, but a negative relationship between land degradation and population density. Accordingly, in the most recent GLADIS study (2010), land degradation was perceived as a complex process, the assessment of which needs to combine biophysical and socioeconomic indicators; hence, the authors developed six axes with several variables included for biomass, soil, water, biodiversity, economics, and social and cultural indicators. Depending on the combination of variables considered, various maps were developed showing, for example, the Ecosystem Service Status or the Land Degradation Impact Index. However, how to interpret these complex maps remains an issue. In addition, some findings at the global scale could not be confirmed at the national or regional level. For example, a regional approach on Sub-Saharan Africa conducted by Vlek, Le, and Tamene (2008, 2010), which followed a different approach toward assessment (including atmospheric fertilization), came to contradictory results. Their results appear to be more applicable to Sub-Saharan Africa. However, the LADA/GLADA study did raise awareness regarding the problem of land degradation, in particular in humid areas, as they found that 78 percent of the areas affected by land degradation are located in humid areas (Bai et al. 2008b).

The extent of land degradation, soil degradation, and desertification identified by all of these studies varies and is hardly comparable between studies. GLASOD, for example, estimated that 65 percent of the land is degraded to some extent, whereas GLADA considers 24 percent of the land area to be degraded between 1981–2003. For a global assessment of land degradation, remote sensing and georeferenced data are definitely needed; however, results still have to be validated on the ground before they are considered reliable.

Acknowledging the work of GLADIS, a more systematic choice of socioeconomic data needs to be developed. Cost calculations of global land degradation are urgently needed to be compared to the actual cost of action against it.

Why does land degradation occur? Proximate causes are direct drivers of land degradation. These causes include biophysical factors (climatic conditions, topography) and unsustainable land management techniques (clearing, over-grazing, and so on). Underlying causes comprise policies and institutional and other socioeconomic factors that have an impact on the proximate causes. In addition, proximate and underlying causes may be related to each other, which makes it difficult to assess the influence of a single factor. National, international, and local policies and strategies; access to markets; infrastructure; the presence of agricultural extension services; population density; poverty; and land tenure conditions were all empirically proven to matter. Of course, other possible factors also exist, depending on the specific local conditions.

Empirical results done at global level to illustrate our approaches to analyzing the association between change of land cover and key biophysical and socio-economic variables show a strong global associations between some of explanatory variables and changes in NDVI. Some of these relationships are consistent across different geographical areas (agricultural intensification and government effectiveness) while other show complex differences by regions (population density and economic growth). While the results for government effectiveness are not surprising, our estimates about the effects of agricultural intensification on NDVI and possibly on land degradation are important and not expected. This is particularly true because our analysis, by excluding most of agricultural areas, applies mostly to natural vegetation and areas with very little agriculture and most likely of subsistence nature. The significance of this results calls for additional scrutiny. We do not find clear evidence of the *more people less degradation* hypothesis as the signs for the parameters change across regions. Similarly the effect of economic growth is not univocal. In fact, these two last results suggest that land and soil degradation have strong regional and national dimensions and as such need to be analyzed. More complex analytically sound modeling is called for to capture the drivers and guide global, regional, and national policies.

Land degradation has many impacts on the environment, the economy, and society. Many of these effects are externalities, meaning they are not transmitted through prices and, hence, not considered in an individual's land use decision. In the literature related to land degradation, those effects are usually classified into on- and off-site effects. On-site effects of land degradation describe the impacts that can be directly experienced by farmers, such as declining yields. Off-site effects—as externalities—are effects that do not occur on the degrading land itself. Sedimentation due to soil erosion, for example, can lead to siltation of reservoirs and dams, with negative impacts on navigation and reservoir storage capacity for irrigation, domestic water supply, industries, and hydropower. It may also lead to disruption of the stream ecosystem and a reduced value of recreational activities at those sites. Wind erosion can lead to dust storms, which have negative impacts on human health and which increase cleaning and maintenance costs.

Another class of effects that arise due to externalities are so-called indirect effects. Land degradation affects agricultural production and incomes, thus affecting the prices of inputs and of goods produced. Further, the impacts on the agricultural market have intersectoral, economywide (multiplier) effects and may lead to food insecurity, poverty, migration, and other outcomes affecting the society. A global assessment must consider all these relevant effects of land degradation in order to come up with adequate estimates of the total cost of LD.

3. Economics of LD

Land degradation matters to people, because it affects

- the range of activities that people can undertake on the land and the range of services provided by the land—in other words, it restrains choices and options;
- the productivity of these activities and services and, thereby, the economic returns they generate; and
- the intrinsic or existence value of the land.[30]

Degradation affects the economic value of land, because this value is based on its capacity to provide services. These services include not only physical output (for example, food and resource production) but also other services beneficial to human well-being (for example, recreational parks). The existence of ethical, philosophical, and cultural considerations that give ecosystems a value—irrespective of their benefits to humans—will not be part of this economic assessment. We acknowledge that these considerations exist but believe they are better addressed by societies using other processes than economic analysis (Pagiola, von Ritter, and Bishop 2004).

To date, the most studied impact of land degradation (LD) is the decline in crop yields. In this report, we recommend also considering the production of a wider, more comprehensive range of services in land ecosystems. Some of these services are valuable for their support to agricultural systems (regulation of water supplies for irrigation, pollination, genetic resources for crop improvement, and so on) but can also provide services that go beyond agricultural production (for example, carbon sequestration, flood control, recreational activities.). The Millennium Ecosystem Assessment (MA 2005a, 2005b) and The Economics of Ecosystems and Biodiversity (TEEB 2010) identified three types of benefits derived from ecosystem services that are affecting human well-being—ecological, sociocultural, and economic benefits—all of which can be affected by LD.

Most of the studies on the economic valuation of ecosystem services focus either on a selection of benefits arising from a particular service or on ecosystems at a specific location. Few studies have attempted to estimate the value of the full range of services by regions or for the whole planet (Pagiola, von Ritter, and Bishop 2004). Pimentel et al. (1997), in analyzing waste recycling, soil formation,[31] nitrogen fixation, bioremediation of chemical

---

30 As mentioned in the objectives in Section 1, the intrinsic value of land resources is not covered in this report.

31 Based on Pimentel et al. (1995), a conservative total value of soil biota activity to soil formation on U.S. agricultural land is approximately $5 billion per year. For the 4.5 billion

pollution, biotechnology (genetic resources), biological pest control, pollination, and the support of wild animals and ecotourism, estimated that the worldwide economic value of these services is $2,928 billion, of which 49 percent is due to waste recycling alone. Costanza et al. (1997) estimated an average annual value of nature's services of the entire biosphere to be $33 trillion per year, which is more than the world economy of $18 trillion per year. A review of studies on ecosystem valuation performed by Balmford et al. (2002) returned a mean total cost of a global reserve program, on both land and sea, of some $45 billion per year.

Some of the existing studies on the costs of land degradation express the costs as a share of gross domestic product (GDP).[32] But how can the impact of LD on present and future well-being (not only economically) be measured as a loss in GDP (that is, the amount of goods and services produced in a year and in a country)? Of particular importance is a work commissioned by French president Nicolas Sarkozy and written by the Commission on the Measurement of Economic Performance and Social Progress (CMEPSP). This study addressed the problems with existing measurements of well-being and stressed the need for reliable indicators of social progress (Stiglitz, Sen, and Fitoussi 2009). Despite its widespread use, GPD does not provide information about the well-being of a nation's social and environmental factors. The commission recommended renewing the efforts to develop the GDP measure beyond the narrow focus on productivity to account for health, education, security, environment, and sustainability. This report is therefore in line with TEEB (2010, MA (2005a), and Stern (2006), which also advocate the use of a method that takes into account all economic, social, and environmental costs and benefits.

The commission's suggested approach to measure sustainability is of particular relevance for the economics of LD (ELD). Two indicators are of particular interest: adjusted net savings and ecological footprint. The former attempts an assessment of the economic component of sustainability, which means keeping a constant stock of extended wealth in a country—with extended wealth comprising natural resources, physical capital, productive capital, and human capital. This approach is reasonable for items that can be assessed using existing economic valuation techniques. However, the concept fails to account for the global nature of sustainability and has a limited applicability for the many environmental goods for which constructing monetary values is still difficult. The commission advised that the "economic" sustainability measure should be complemented by a set of well-chosen physical indicators, which could focus on aspects of environmental goods that remain difficult to measure in monetary terms. It is in this framework that the ecological footprint measure

---

hectares of world agricultural land, soil biota contribute approximately $25 billion per year in topsoil value (Pimentel 1997).

32 Examples are provided in Appendix 5.

is used. This indicator focuses exclusively on natural resources by calculating the amount of land and water required to maintain a given level of consumption.

Aglietta (2010) also tried to address the issue of sustainability and provided a framework in which wealth accounting and social welfare under sustainability are connected. All assets contributing to economic welfare are capitalized. These assets include public services that are produced by tangible and intangible assets[33] owned by society as a whole. The different forms and definitions of capital and assets are depicted in Figure 3.1. It must be noted that the adoption of this approach is hindered by missing data and comprehensive measurement of the state of capital and assets. More effort and coordination must be undertaken to make this concept workable.

Figure 3.1—Total wealth and social welfare

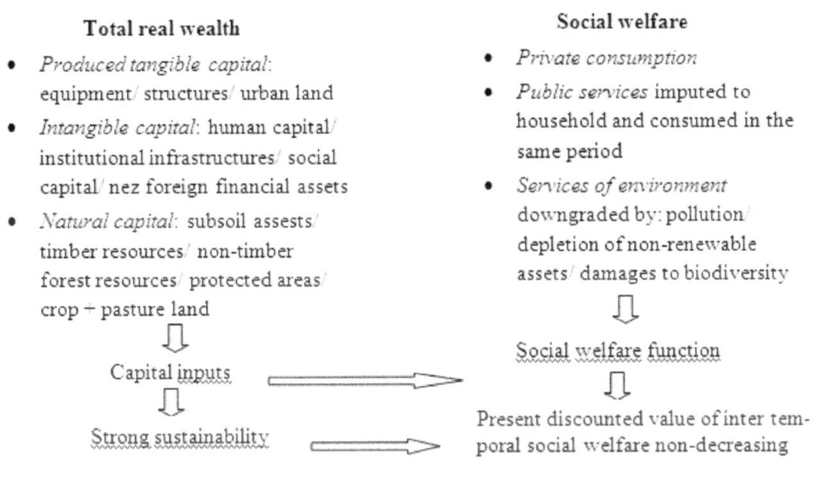

Source: Based on Aglietta 2010.

33 The term intangible asset describes an asset that is not physical in nature. Corporate intellectual property (such as patents, trademarks, copyrights, business methodologies), brand name, long-established customers, and exclusive supplier agreements are common examples of intangible assets (see Cohen 2005).

3.1 Theoretical Framework for the Economic Valuation of LD Impacts

This section first presents background information necessary for the economic valuation of land resources. A methodological framework is developed so that a systematic comparison between the costs of action against land degradation and the costs of inaction is possible.

Natural resources are often classified as either nonrenewable or renewable. Land is considered to be in between these two categories and is treated as a semirenewable resource. When the rate of depletion is faster than the rate of regeneration, the land resource is degraded. The actual rate of land degradation depends on many factors—some of which are site specific, such as soil type, slope, and climate, whereas others are dependent on land user's choices (for example, production technology and cropping systems). It is often the case that degradation rates for agricultural land exceed naturally occurring rates (Barbier 1999).

Land is a fundamental input in agricultural production, and fertility is one of its most important characteristics. Considerations about land productivity and land degradation are implicitly or explicitly incorporated in the farmers' decision processes. Although conservation measures often do not erase all the negative effects, they are often capable of mitigating the consequences of degradation. For instance, depending on the degree of substitutability between human-made capital and natural capital, one can restore fertility by increasing the use of inputs, by changing land management, or by changing the cropping system.

Actions against degradation are beneficial for the land but usually lead to higher production costs for farmers (in terms of labor or capital requirements or lost productive area). Economic analysis helps address the question arising from these trade-offs, such as whether the benefits due to soil conservation are worth the additional costs (Lutz, Pagiola, and Reiche 1994; Requier-Desjardins 2006).

The economic assessment of environmental and climatic problems has received increasing international attention in recent years. The Stern Review on the Economics of Climate Change was released for the British government in October 2006 (Stern 2006). TEEB was launched as a consequence of the G8+5 Environmental Ministers Meeting in Potsdam, Germany, in March 2007. The main outputs of TEEB were an interim report released in May 2008 describing the first phase, and the final reports are targeted at specific end users (policymakers, businesses, administrators, consumers) of the second phase (2008–2010). More details on the Stern Review and TEEB can be found in Box 3.1.

We propose following a framework similar to that put forward by those reports: an economic evaluation of the costs of action (that is, the costs of mitigating land degradation) versus the costs of inaction (that is, the costs induced by continued degradation). Because land degradation is a process that takes place over time, intertemporal considerations will characterize farmers' decisions. This means that the benefits derived from land use (and the value of the land) need to be maximized over time and that farmers must continuously choose between land-degrading and land-conserving practices. From an economic perspective, the current profits of adopting land-degrading practices are continuously compared with the future benefits that derive from the adoption of land conservation practices. A rational farmer will let degradation take place until the benefits from adopting a conservation practice equal the costs of letting additional degradation occur. Each farmer determines his or her own optimal private rate of land degradation.[34] This optimal private rate mainly depends on the costs and benefits that the farmer directly experiences, such as yield declines due to degradation. Typically, productivity losses are referred to as *on-site costs* (taking place on the farmer's plot of land). Hence, those ecosystem services that result in lower production levels are considered in the decision, whereas those that do not become measurable in terms of lost production are neglected. In fact, many of the costs related to land degradation do not directly affect an individual farmer. As a consequence, the private rate of degradation is not likely to reflect the optimal rate of degradation from society's viewpoint.

Box 3.1—Recent major economic assessments of the environment

Stern Review

The Stern Review deals with the economics of climate change as a result of externalities from greenhouse gas emissions. Climate change leads to global consequences (off-site effects) taking place over a long period and its analysis involves ethical dimensions, because the impacts of climate change are not equally distributed among countries, people, and generations. The economic analysis needs to take into account that impacts of climate change are long term and persistent, even irreversible, and are associated with uncertainties and risks. Results are therefore dependent on assumptions about plausible future emission scenarios, as well as on assumptions about technical progress and discount rates.

The review has focused on the costs of mitigation to reach the stabilization of greenhouse gas concentrations in the atmosphere in the range of 450–550 parts per million of carbon dioxide, whereas the inaction path ("business as usual") is associated with a temperature increase of 2–3 degrees Celsius. Costs of inaction are then estimated at an average reduction of at least 5 percent in global per capita consumption. Costs of mitigation are estimated at around 1 percent of the global GDP by 2050 on average, with a range of –2 to +5 percent of GDP. Comparison of the costs of mitigation with the costs of inaction suggest that there is a net gain in taking action to mitigate climate change now rather than bearing its consequences. The social costs of carbon are taken as $85 per ton of carbon dioxide, which is well above the marginal abatement costs in many sectors. The net present

34 The optimal rate of degradation will thus not lead to zero degradation but will usually include at least some level of land degradation.

Box 3.1—Continued

value suggests net benefits in the order of $2.5 trillion when implementing a strong mitigation policy in 2011. There is a high price to delay taking action on climate change. As these findings show that strong action on climate change is beneficial, the second half of the report examines the appropriate form of such policy and how to fit it in a collective action framework.

TEEB

TEEB provides ways to create a valuation framework for ecosystems and biodiversity in order to address the true economic value of ecosystem services by offering economic tools. The analytical framework of a cost–benefit analysis must deal with the economics of risks and uncertainty. Crucial for any cost–benefit analysis is the setting of discount rates to make future losses and benefits comparable. Decisions on discount rates involve ethical dimensions; therefore, TEEB intends to present a range of discounting choices connected to different ethical standpoints. TEEB reviews a number of studies concerning the costs of both biodiversity loss and biodiversity conservation. Monetary values attached to biodiversity and ecosystems have often focused on case studies (area-specific) and on particular aspects of ecosystems or sectors. Assessing the consequences of biodiversity loss and ecosystem services globally thus demands a globally comprehensive and spatially explicit framework and estimation grid for the economic valuation of ecosystems and biodiversity, combined with a meta-analysis of valuation studies. The key elements of Phase 2 of TEEB include the causes of biodiversity loss; the design of appropriate scenarios for the consequences of biodiversity loss; the evaluation of alternative strategies ("actions to conserve") in a cost–benefit framework, including risk and uncertainties; a spatially explicit analysis; and the consideration of the distribution of the impacts of losses and benefits. The evaluation also largely relies on benefit transfer, because data cannot be collected for all kinds of ecosystem services and biomes.

From a society's point of view, all costs and benefits (including externalities) that occur due to ongoing land degradation need to be considered to result in the optimal "social" rate of land degradation. This includes not only on-site and direct costs that farmers experience in terms of lower yields, but also changes in the value of the benefits derived from all ecosystem services that may be affected; off-site costs arising at other sites within the watershed, such as sedimentation; as well as indirect effects, such as economywide impacts, threats to food security, poverty, and other outcomes affecting the society. (See Figure 1.2 for a description of the various costs and how they are linked to externalities and social costs.)

Government policies and other institutional factors can also lead to socially and privately nonoptimal rates of land degradation. Imperfect or unenforced land rights, distorted and volatile market prices, lack of information about future damages related to degradation, and imperfect or missing credit markets are among the factors that prevent farmers from investing in potentially profitable soil conservation measures. Anything that creates uncertainty about the future benefits of conservation measures reduces farmers' incentives to adopt them.

When the costs of land degradation are not paid in full by producers (that is, when marginal social costs are higher than marginal private costs) or when there are misperceptions about the benefits deriving from letting degradation occur, the resulting rate of degradation is higher than socially optimal, and total social welfare is suboptimal.

Figure 3.2 provides a stylized representation of the problem and demonstrates how action toward land degradation can be insufficient from a societal perspective.

Figure 3.2—Marginal private and social costs of land degradation

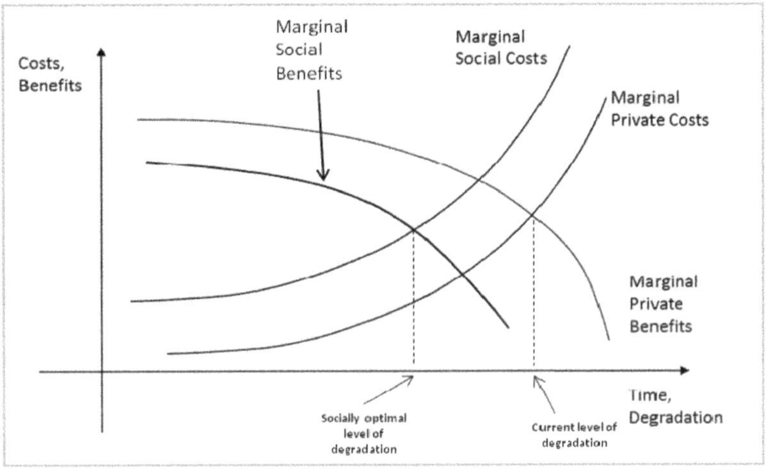

Source: Author's creation.

Cost of Action versus Inaction

A possible way to address this problem is to compare the costs of action against the costs of inaction. Action is meant to include all possible measures that can be taken to avoid or mitigate land degradation or to restore degraded lands. Although these measures normally involve soil and water conservation measures, changes in institutional structures or policies are included as well, as is any appropriate mix of all of them. More simply, inaction describes business-as-usual behavior. This approach can be made operational by comparing marginal costs and benefits (that is, the costs and benefits of an extremely small change in the level of degradation) related to degradation (Figure 3.2), and it is, from a practical point of view, more tractable than other methods. For the application of this method, it is paramount that information about the marginal social cost related to continued degradation (marginal costs of nonaction) and the marginal social cost related to conservation (cost of action) can be gathered.

To construct the marginal cost curves, we first need to develop production functions that link the extent of degradation to the maximum agricultural output associated with a technology (nonconserving or conserving). This allows us to capture the on-site productivity loss as the most direct impact of land degradation on farmers. In addition to direct costs and (at least short-term) benefits of land degradation, off-site costs and benefits, as well as indirect effects, need to be taken into account. To come up with a socially optimal degradation, among the economic valuation methods presented in a coming section, suitable methods have to be identified to address the various on- and off-site and direct and indirect costs and benefits. Time plays a vital role, as the impact of land degradation may aggravate over time; therefore has to be incorporated as well. Costs and benefits that arise over time have to be discounted in order to be comparable. Due to the current lack of knowledge on the long-term impacts of agricultural practices on degradation rates (and potential price fluctuations), uncertainty also has to be incorporated in the analysis.

An appropriate economic tool for a systematic comparison of all costs and benefits (private and social) of continued land-degrading practices and specific land-conserving actions is the cost–benefit analysis (CBA), which can be used to discount costs and benefits to come up with one comparable value. Discounting is a procedure in which costs and benefits are valued less the more distant they are in the future.[35] CBA makes future costs and benefits comparable by using the net present value for investments in conservation measures and continued degradation. The net present value is the discounted net benefit gained or the net cost imposed. Sensitivity analysis allows coping with uncertainty by analyzing the sensitivity of the results obtained under the cost–benefit analysis to variations in the risk factor.

The marginal cost of action curve (often referred to as marginal abatement cost curve, or MAC) consists of various measures, such as soil and water conservation techniques, institutions, and policies, and their cost to abate degradation by one unit. On the MAC, each point along the curve shows the cost of (a combination of) action(s) to abate degradation by one additional marginal unit, given the existing level of degradation. In this case, marginal changes refer to changes in LD caused by a single (or combination of) measure(s). The rising MAC curve (positive slope) indicates that as more abatement has been achieved, the cost of the next unit of abatement increases—that is, the MAC is an increasing function of the level of abatement.

Unfortunately, in practice, the MAC curve is difficult to observe or estimate. One reason for this difficulty is that farmers rarely apply a single

35 Discounting is based on the observation that individuals prefer to enjoy their benefits now and bear the costs later (intergenerational equity issue).

abatement strategy. In addition, the level of degradation is rarely recorded before and after application of a given strategy and the sequence and abatement results of each abatement strategy is rarely recorded.

One way to approximate the MAC curve based on conservation measures is illustrated in Figure 3.3. It must be understood that such a construction of the MAC is not grounded on theory and is a coarse approximation of the real MAC curve. Nonetheless, similar techniques have been successfully applied in other contexts of natural resource conservation to guide policy choices (see McKinsey and Company 2009 for the case of water). As (combinations of) abatement strategies are applied (independently of each other but within a given study region), their impacts on specific processes of degradation (for example, levels of soil nutrient, water retention, or erosion) are measured, controlling for other factors affecting degradation (weather/climate, slope, working practices of the farmers, and so on). Given the number of "units" of degradation that are abated by these measures and given their total cost, an average cost of abatement over the range of abated degradation is computed, albeit in abstraction of how much abatement had already been achieved before the implementation of this specific (combination of) strategy(ies).

In Figure 3.3, each column represents a (combination of) strategy(ies)—the width of the column is its impact on degradation, and the height of the column is its average cost per unit of abated degradation. The strategies depicted on the left indicate negative costs; these are cases in which "doing less brings more"—for example, correcting the current production practices strategies decreases the production costs for a given level of output while improving land conservation. A typical win–win situation occurs when fertilizers have been overused, leading to strongly decreasing marginal returns in yields per unit of fertilizer and causing degradation issues such as salinity and other chemical degradation. The strategies on the right are the conservation strategies, which have a higher average cost per unit of abated degradation. Typically, these can be expected to be strategies, such as terracing, that involve a great deal of labor, equipment, and machines. The horizontal aggregation of average costs over given (small) ranges of degradation abatement can be viewed as an approximation to the MAC curve. The marginal cost of the nonaction curve represents, rather straightforwardly, the continued impact of nonconserving and land mining agricultural (and other land use) practices on costs. In extreme cases, these costs can come to land being abandoned by the farmer (or at least its original or most profitable use is abandoned), in which case the cost of nonaction is equal to the value of the foregone production, net the costs of conservation measures. In most cases, action against degradation is eventually taken. Yet the costs of nonaction rise with time and with increasing levels of land degradation, as illustrated in Figure 3.4. At the beginning of the curves (left), when degradation levels are low, avoiding an additional unit of degradation brings a net benefit (the win–win situation already mentioned). As action is delayed, it can initially return even larger net benefits as early action. Yet this should not be an incentive to delay action, as this logic would be similar to purposely degrading the land now for the sake of higher returns from conservation measures later.

Figure 3.3—Example of a marginal abatement cost curve

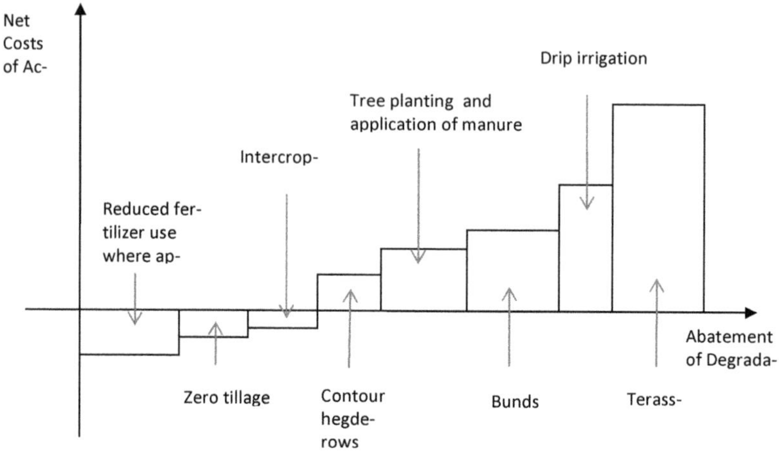

Source: Author's creation.

Nonetheless, in many situations, land users are now facing cases in which action has been delayed for so long that the benefits that can be earned from a win–win situation have become substantial. But the striking feature of the delayed action curve is that past a certain threshold, once win–wins have been exhausted, the cost of further abatement of degradation rises sharply above the cost of earlier action, due to continuous degradation increases, which cause higher productivity losses and negative off-site effects. As action is delayed, it becomes more difficult to restore already-lost productivity and to mitigate continued negative off-site effects. A further impact of delayed action is that the price of land increases (P0 to P1), because nondegraded land has become scarcer. The price of the land determines the amount of action that is economically optimal to undertake in order to limit or reverse degradation. For delayed action, it is obvious from Figure 3.4 that the costs of optimal abatement are much larger (at P0D0) than if early intervention is chosen (at P1D1). Figure 3.4 is only an example, and the curves it portrays can be expected to vary greatly in their shape according to each specific situation. Nonetheless, it is this sort of curve that, once developed, will help guide policy action against LD. Even when the curves are approximated, as in Figure 3.3, they can help identify the crucial features of the problem at hand—that is, how much effort should be devoted to fight LD, where, and at which cost (in Figure 3.4, approximated by the difference between P0 and P1).

Figure 3.4—Cost of action and cost of delayed action

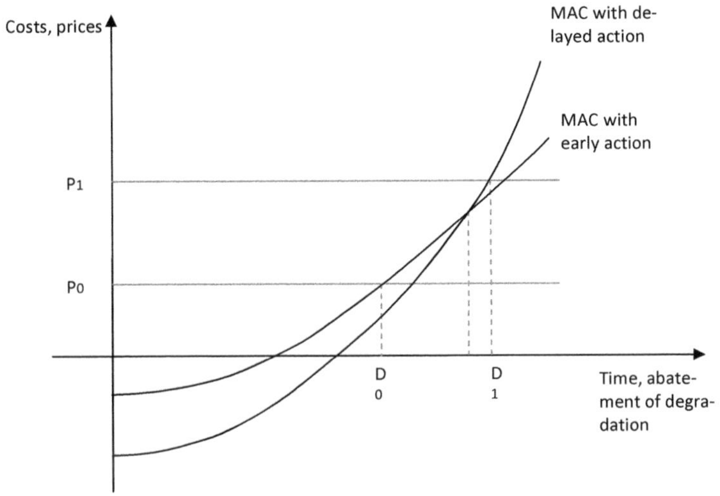

Source: Author's creation.

This framework, which can be implemented at reasonable data costs, should be undertaken in several representative areas, thus bringing the site specificities of LD into its global economic assessment. So far, valuation studies of the costs and benefits of land degradation and land improvement (as will be presented in the next section) have focused on agroecosystems and their provisioning services. This framework needs to be further developed, in combination with knowledge built into projects such as TEEB, to cover more terrestrial ecosystem services (not only those related to an agricultural output) and their benefits. The approach should be as comprehensive as current science and knowledge allow and should include all the services affected directly or indirectly by LD, which can be achieved similarly to the analysis of agroecosystem services by relying on representative case studies. The case studies must be representative of different ecologies, livelihoods, and institutional settings. Thus, in order to have statistically valid results, the case studies need to be drawn from a global sampling frame. As a second component of the global coverage of land degradation, a global assessment must go beyond case studies to incorporate the transboundary dimensions of land degradation. These studies can be performed at different scales, from localized (for example, erosion in Country A causes sedimentation of dams in Country B) to global (for example, land degradation in a specific area has impacts on global climate or on global food prices). Such transboundary effects of LD must be observed, recorded, and then accounted for through integrated (that is, geographically and sector-wise connected) and dynamic (accounting for the time dimension) modeling approaches.

A Brief Review of Cost–Benefit Analysis Applications to Land Degradation Issues

CBA is a tool suitable for comparing land-degrading and land-conserving management practices over time. It requires knowledge of all costs and benefits associated with practices leading to degradation, as well as of those leading to conservation. The distribution of costs and benefits over time is accounted for by using appropriate discount rates to determine streams of discounted costs and benefits. Common indicators of economic returns are the net present values (NPVs)[36] and the internal rate of return (IRR),[37] which are used to compare alternative scenarios—in this case, adoption or no adoption of conservation practices.

The analyst's choice of the value for the discount rate and time horizon has a crucial impact on profitability and, thus, the results of CBA. Usually, discount rates differ depending on whether they refer to an individual or to the society as a whole. When CBA is applied to evaluate options from an individual's perspective, discount rates are higher, as individuals are thought to have a higher time preference. This assumption is related to their attitudes toward risk and uncertainty, market distortions, and other institutional settings. From a society's perspective, the use of low discount rates is justified with considerations on intergenerational equity and sustainability. Some authors use the current long-term rate of interest, provided by financial markets, as the appropriate discount rate (Crosson 1998). Studies have used discount rates ranging from 1 to 20 percent and time horizons from 5 to 100 years. A useful review is provided in Clark (1996). It is often noted (see, for example, Barbier 1998) that CBA is a difficult implementation due to the required amount of information on costs and benefits.[38] When data are available, CBA can be

36 Present value of cash inflows and cash outflows C; NPV is used to analyze the profitability of an investment. Index t represents the time dimension of the investment, and r is the discount rate: $NPV = \sum_{t=1}^{T} \frac{C_t}{(1+r)^t} - C_0$.

37 The IRR of an investment is the interest rate at which the NPV of the costs equals the NPV of the benefits of the investment. The higher a project's internal rate of return, the more desirable it is to undertake the project.

38 In its basic form, CBA does not consider the distribution of benefits and costs over individuals, and any increase in net benefits is desirable, regardless of to whom they occur (Barbier, Markandya, & Pearce 1998). Considerations about intergenerational equity are

specified to assess the profitability of the adoption of conservation measures. Profits or returns to conservation measures are calculated as the difference between the value of the crop yields and the costs of production.[39] Off-site and indirect costs also need to be subtracted from profit when the CBA is applied from society's perspective. The same procedure can be applied to nonconserving agricultural practices. The NPV of returns to conservation is therefore the difference between the discounted stream of profits with and without the implementation of conservation measures. This method usually estimates the returns to specific conservation measures, not to conservation per se (Lutz, Pagiola, and Reiche 1994).

Numerous studies have applied CBA to analyze the profitability of conservation practices in different areas of the world (see Appendix 2). Whether conservation appeared to be profitable depended on the regions studied and the type of conservation measured (Lutz, Pagiola, and Reiche 1994), the type of crops, the choice of discount rate (Shiferaw and Holden 2001), and the intensity of a conservation measure (for example, hedgerow intensity) (Shively 1999). Nkonya et al. (2008b) found that sustainable land management practices can be profitable from a private perspective as well as from society's perspective.

The decision to adopt conservation measures is based not only on economic reasons (such as costs and benefits) but also on a variety of other factors (Drake, Bergstrom, and Svedsater 1999). A number of studies that investigate the determinants of the adoption of soil and water conservation measures, other than primarily financial factors, are reviewed in a later section on adoption models.

3.2 Economic Valuation of Nature and Its Services—A Review

A full accounting of all the benefits deriving from the various ecosystem services in accordance to MA (2005b) is difficult to undertake. In this report, we focus on the economic benefits. This approach is based on a utilitarian philosophy, and it tends to underestimate the full costs. The utilitarian approach relies on the concept of utility as a measure of value: Alternative states are compared according to the utility they generate, irrespective of how this utility is derived. Environmental benefits are treated equally to any other benefit, assuming they can be traded off. Usually, this means that a loss in

---

not part of CBA analysis, even when the costs and benefits of disadvantaged or poor population groups can be of greater importance than those of better-off groups.

39 Costs are usually related to additional labor and capital requirements for the conservation measure and the loss in productive area. Installing soil conservation measures leads to a reduction of the planted area, which is measured as a loss of crop yield. Additional labor, sometimes capital, is needed to install and maintain conservation measures. Assumptions on labor inputs and wage rates may influence the outcome of CBA; therefore, it is important to obtain an accurate assessment of the costs of labor and capital.

environmental quality can be compensated for by an increase in another type of benefit—for example, income. This does not mean, however, that nonutilitarian considerations regarding land degradation are not relevant;[40] rather, as mentioned earlier, focusing on values sheds light on the economic rationale for conserving land ecosystems.

Decisions about ecosystems and land management typically involve various types of externalities and market failures, since ecosystem services are generally public goods and are not traded in markets. According to Freeman (2003), externalities arise when there is no requirement or incentive for an agent to take into account the effects on others when making choices, which usually happens when these effects are not transmitted through prices. For example, the sedimentation of waterways negatively affects navigation (by causing higher dredging costs). In most cases, the waterway is a public good whose access is not priced or restricted in any manner. Thus, the negative externality (in this case, an off-site cost) imposed on navigators is usually not compensated for by the land users who contributed to the sedimentation.

The role of economic valuation is to place a monetary value on the various ecosystem services provided by the land, even when they are not marketed, so that recommendations regarding the economic efficiency of land management choices can be made. For instance, appropriately pricing the externality imposed on navigators is dependent on their valuation of waterway services. These are the first steps necessary to address the market failure (in this, case a missing market for waterways) and to ensure that the cost of the externality (increased dredging) is internalized in the land management decisions.

A well-established concept for measuring the economic values of natural assets is that of total economic value[41] (TEV, see Figure 3.5). TEV uses multiple values that can be classified according to whether they derive from using the resource (use values) or they are independent from its use (nonuse values). Use values of ecosystem services can be further split up into direct use values (agricultural production, wood, livestock, and so on) and indirect use values (pollination, water purification, and so on). Option value is the value that people place on having the option to enjoy something in the future and not currently use

40 As many have pointed out (see, for example, Randall 1999), there might be alternative methods of ranking decisions and preferences, such as some universally recognized moral imperatives. Therefore, some natural resources, environments, or biological systems should be protected, regardless of the cost and work constraints. Although a cost–benefit analysis may be acceptable as a decision rule in many situations, there can be instances in which societies would expect that decisions about the use of natural resources are based on nonutilitarian principles.

41 "The economic value of a resource-environment system as an asset is the sum of the discounted present values of all the services" (Freeman 2003, p.5).

it. Bequest value is the value that people place on knowing that future generations will have the option to enjoy something (including altruism). Nonuse values are placed on simply knowing that something exists, even if people will never see it or use it. In most studies, the estimated values consist of (direct) use values because they are related to the mainly agricultural use of the land (Requier-Desjardins 2006). Yet, many of the services provided by the land are not traded on markets and thus require nonmarket valuation techniques (Freeman 2003). Only a small number of studies have attempted to measure indirect values or nonuse values. The Österreichische Bundesforste AG (2009) attempts to value biodiversity, climate change mitigation, soil stabilization, and cultural services provided by protected areas in Ethiopia. The results suggest that the value of biodiversity in developing countries is in the range of $1–$30 per hectare per year. Schuyt (2005) discussed several economic valuation studies carried out for different African wetlands. In this study, the existence value of biodiversity was measured using the contingent valuation method and was estimated at $4,229,309. Ecosystems that provide multiple services or that are of a regional importance for other, dependent ecosystems provide benefits that may reach an order of magnitude that is equal to or larger than the direct use value of the ecosystem in question (Görlach, Landgrebe-Trinkunaite, and Interwies 2004).

Figure 3.5—Total economic value framework

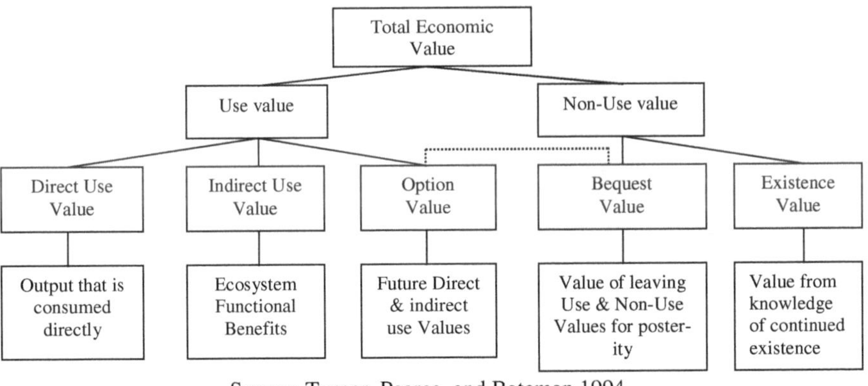

Source: Turner, Pearce, and Bateman 1994.

The approach provided by MA (2005b), as presented in Box 1.1, can be linked to the TEV concept as follows, though this is not straightforward. In general, direct use values broadly match provisioning and cultural services, whereas indirect use values match regulating services, and existence value partly overlaps with cultural services.

There is no particular overlap between supporting services and any value within the TEV concept; they are valued implicitly because they are essential for the functioning of the ecosystem and, hence, provision services (Pagiola, von Ritter, and Bishop 2004).

Balmford et al. (2008) stated that the approach of MA (2005) mixes benefits obtained from ecosystem services and the processes by which these benefits are delivered. Mixing processes and benefits easily leads to double-counting of values, resulting in an overstated value of ecosystems. Balmford et al. (2008) provided an example to demonstrate this problem: Water purification could be valued as both a regulating service and a benefit (from drinking water). If the value of the service is quantified in addition to the value of fresh water, there is clear double-counting in the valuation of water purification. As a possible remedy Balmford et al. (2008)[42] made a distinction between processes and benefits (see Figure 3.6). Processes refer to core ecosystem processes, which consist of the basic ecosystem functions (for example, nutrient cycling, water cycling) and beneficial ecosystem processes (such as waste assimilation, water purification) that directly lead to benefits for humans (such as clean drinking water). These benefits can then be valued in monetary terms. Focusing solely on the valuation of the benefits derived from ecosystem services is the key to avoiding double-counting.

However, the authors were aware that—despite its theoretical appeal—this framework for economic valuation has to be adapted to practical considerations (such as the focus of previous studies and, hence, available data and information). In the end, the authors resorted to looking at ecosystem services within thematic groups that are still a mix of benefits and processes. Their work illustrates the difficulty of operationalizing any framework that aims at valuing all the benefits that humans derive from nature. Economists are continuously improving methodologies to address this issue. At this time, we suggest relying on the TEV and classifications of ecosystem services and benefits, such as that presented in Balmford et al. (2008), in order to avoid the double-counting issue.

42 Supporting services, as suggested by MA (2005), correspond broadly to the core ecosystem services, including some beneficial services. Regulating services mirror "beneficial" processes, and provisional and cultural services are reflected by "benefits" (Balmford et al. 2008).

Figure 3.6—Ecosystem services framework

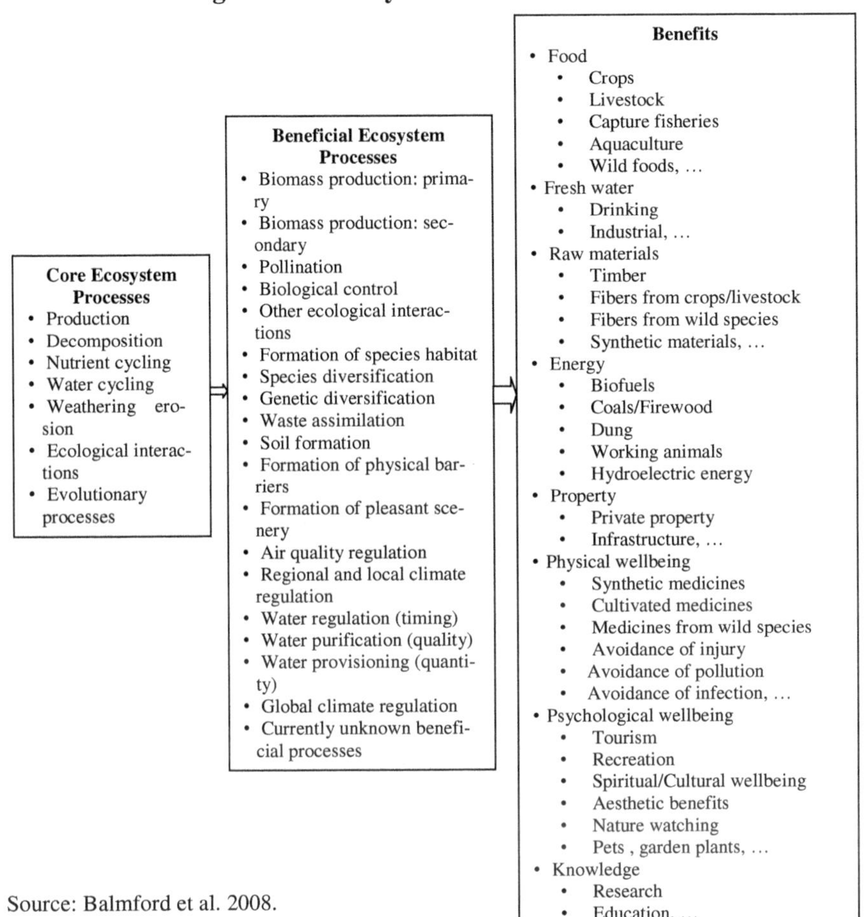

Source: Balmford et al. 2008.

Valuation Techniques for the Costs and Benefits of Ecosystem Services

A wide range of valuation approaches can be used to assess the value of natural resources.[43] The main ones are reviewed in Table 3.1. To infer the costs and benefits of land degradation and land improvements, studies in the literature mainly employ the following:

43 See, for example, Requier-Desjardins, Adhikari, and Sperlich (2010) for an overview of economic ecosystem valuation techniques or Görlach, Landgrebe-Trinkunaite, and Interwies (2004).

1. Replacement cost approaches
2. Nonmarket valuation techniques, such as hedonic pricing, contingent valuation, and choice experiments
3. Productivity change approaches
4. Avertive behavior and damage costs approaches
5. Benefit transfer

The methods under (2) and (4) are of special importance in the valuation of off-site costs of land degradation. Values obtained with any of the methods can theoretically be transferred to other sites when the valuation context is similar by using the benefit transfer method.

Table 3.1—Main valuation techniques

Methodology	Approach	Limitations
Market price	People's actual WTP	Missing or distorted markets
Production function–based approaches/ Productivity change	Economic contribution of ecosystems to production of goods and services	Relationship between change in ecosystem services and production level
Travel cost	Spending to access ecosystem sites	Limited to recreational benefits
Hedonic pricing	Difference in prices (of property, wage) due to existence or level of ecosystem services	Information on ecosystem services may not be transparent; distorted markets
Replacement costs	Costs of replacing ecosystem services and goods	Assumption that artificial replacement is equivalent
Mitigative or avertive behavior	Costs of mitigating or averting the loss of ecosystem services	Prone to overestimation
Damage cost avoided	Avoided costs of land degradation	Avoided costs may not be equal to benefits of ecosystem services
Contingent valuation	The amount people are willing to pay or to accept	Potential biases; context specific
Choice experiment	Choice of a preferred option from a set of alternatives with particular attributes	Potential biases; context specific
Benefit transfer	Results obtained in a specific context are transferred to another comparable site	Can be inaccurate; depends on how similar context or factors are

Source: Modified based on Requier-Desjardins, Adhikari, and Sperlich 2010.
Note: WTP = willingness to pay.

However, benefit transfer should be used with caution, as it can be misleading when important influencing factors are underestimated or unknown.

Replacement Cost Approaches

The replacement cost approach helps calculate the costs of restoring land's capability of providing ecosystem services after LD. In most studies, this approach is used to value the impacts of soil erosion (as one of the processes of land degradation)—in particular, nutrient depletion—by calculating the costs associated with the application of chemical fertilizer to replace the lost nutrients.[44] The method is often criticized for a number of reasons. First, studies fail to consider the full impacts of land degradation, because the method focuses on one aspect of LD (usually soil erosion processes), and often does not account for the damage caused by other aspects of erosion to soil characteristics, such as organic matter content and physical structure. The method assumes that a perfect substitute for the loss in ecosystem functioning exists that allows for the provision of the same level of benefits as before. However, adding chemical fertilizer is usually insufficient to fully restore soil functionalities, and soil nutrient reserves in particular are ignored; inefficiencies of fertilizer due to leaching and vaporization need to be taken into account (Jayasuriya 2003). The actual replacement with quantities of artificial fertilizer would lead to negative off-site effects. In order to calculate replacement costs, all costs associated with replenishing nutrients, including transportation, labor, and energy costs, need to be considered. As a result, the replacement cost approach is not helpful for the selection of the most appropriate conservation action. This method is likely to overestimate the values of soil nutrients, as it does not establish any connection between soil nutrients and agricultural production. A decrease in nutrients may have little effect on production, especially when other factors, such as rainfall, represent more important production constraints (Bojö 1996; Lutz, Pagiola, and Reiche 1994).[45] Widely cited applications of the replacement cost approach were developed by Stocking (1986) in Zimbabwe and by Stoorvogel and Smaling (1990), who estimated the nutrients budgets of all Sub-Saharan African countries (Bojö 1996). Pimentel et al. (1995) considered (in addition to wind and water erosion) soil depth, biota, organic matter, and water resources in their analysis and included costs related to the energy requirements to replace lost water and the application of fertilizers. They scaled up the cost estimates for the United States in order to give an estimation of worldwide costs of soil erosion, which came out to $400 billion per year.

44 Clark (1996) combined the replacement cost approach and the valuation of physical soil loss by calculating costs of the return of eroded sediments into one of the two main approaches on valuing soil erosion in order to estimate the impact on the soil's properties.

45 Barbier (1998) criticized the method because it compares nutrient loss to a situation without degradation, though zero degradation is practically infeasible.

Nonmarket Approaches

Hedonic Pricing

Hedonic pricing uses realized market prices to infer how much people value changes in the attributes of the goods sold. With well-functioning land markets, the price of land can be assumed to be equal to the sum of the appropriately discounted stream of net benefits derived from its use (Freeman 2003). Hedonic pricing assumes that differences in property values are attributable to—controlling for other things—different levels of land degradation (Jayasuriya 2003). As King and Sinden (1988) pointed out, well-functioning land markets are not always featured in developing countries. Because market prices implicitly reflect the buyers' knowledge of costs and benefits related to the land's productive capacity, the method tends to underestimate the costs of degradation, especially in the presence of off-site costs (Bishop 1995).

Contingent Valuation

Contingent valuation is a survey-based method to determine a monetary value of nonmarketed goods. An individual's willingness to pay (accept) is measured in a hypothetical market scenario as a stated amount of money (price, entrance fees, taxes, meals, and so on) that a person would be willing to pay (accept) for an increase (decrease) in the provision of the good (Freeman 2003). The monetary values obtained are contingent on the hypothetical market scenario and the described resource. The method is known to be associated with a number of possible biases,[46] which lead to either under- or overestimation of willingness to pay. Contingent valuation, as well as choice experiments, is a suitable method to value off-site effects related to land degradation. (A review of studies applying hedonic pricing, contingent valuation, or choice experiments is provided in a later section.)

Choice Experiments

Choice experiments, another survey-based method, ask an individual to choose the most preferred option or alternative from a set of proposed options. These options differ in their characteristics or attributes. Attributes are selected so that they meaningfully describe differences between options in order to explain preferences. Each attribute consists of a set of levels to represent variations in the respective attribute among the options. Attributes and levels are combined into options according to statistical design principles (Louviere et al. 2000). Choice experiments can detect the relative importance of the different attributes and can identify willingness to pay for single attribute changes as well as for aggregate benefits of different policy scenarios.

46 See for example Mitchell and Carson (1989) for a discussion of the relevance of various possible biases.

Productivity Change Approaches

The productivity change approach is the most commonly used method to value land degradation, particularly when looking at soil erosion. This method is based on the idea that a value can be placed on the services the land provides—usually, the agricultural output that it can generate. This approach assumes that all impacts of land degradation manifest themselves through a loss of agricultural productivity. Therefore, land is valued in terms of lost production, sometimes termed the *production equivalent of degradation* (the value of foregone production). Studies typically measure the physical effects of soil erosion, salinity, and soil compaction on crop yields and productivity, though rarely are all the impacts of all the processes that land degradation comprises analyzed. The choice of an appropriate benchmark, or baseline, against which changes are compared is of fundamental importance. An appropriate benchmark is to compare the costs of land degradation to the costs and benefits of actions against it.

Even though implementation of the productivity change approach is relatively straightforward, the method has its shortcomings. Crop prices may be poor indicators of value when markets are poorly developed or distorted by agricultural policies (Crosson 1998). It is also often difficult to account for farmers' reactions to degrading soil characteristics. Since farmers are likely to adopt a mix of inputs to offset damages caused by erosion, it might take some years before degradation manifests itself in the form of yield declines.

Box 3.2 gives an overview of methods used to quantify the extent of land degradation (mostly measured as erosion). Linking agricultural yields and productivity to land degradation is a difficult task. Crop yields, however, depend not only on land degradation but also on a variety of factors, such as management practices, choice of crops, climatic factors, pests, and diseases. Agricultural productivity is the result of the dynamic interaction of numerous factors, and thus it is difficult to disentangle the effect on crop yield related to land degradation only (Lal 1987). Isolating the effects of a single factor, such as erosion, on crop yields represents a real challenge. The impact of soil erosion on productivity can be estimated econometrically or through biophysical models that simulate the interaction between biophysical factors on productivity. The Erosion Productivity Impact Calculator (EPIC), developed by Williams, Renard, and Dyke (1983), is an example of a model that generates erosion rates and the resulting loss of crop yields, given farm management practices. The parameters and coefficients were fitted to conditions in the United States; therefore, this model has to be adapted for use under other local conditions, requiring a great deal of data. Aune and Lal (1995) developed the Tropical Soil Productivity

Calculator for special conditions in the tropics.[47] There have been various attempts to econometrically measure the soil erosion–productivity relationship. A common solution is to establish a direct relationship by using simplified yield functions that have topsoil depth as a dependent variable[48] (Gunatilake and Vieth 2000). Key equations that link the economic behavioral model with the biophysical system of land degradation are production functions that include the effect of changes in soils due to land degradation. Many existing studies have estimated the impact of conservation measures on productivity and have compared it to the impact of degrading (nonconserving) agricultural practice.

Box 3.2—Measuring land degradation

Determination of soil erosion rates provides the basis for analysis in the majority of the studies (Enters 1998). According to Oldeman (1996), erosion is the most important driver of land degradation. As such, a variety of methods and models exists to quantify the extent of degradation by determining soil erosion. A widely used approach to predicting soil erosion is the Universal Soil Loss Equation (USLE; Wischmeier and Smith 1978). The equation predicts mean annual soil loss from various variables, such as the erosivity of rainfall, the erodibility of the soil, the length and slope of the soil, crop cover and management factors, and a conservation practices factor. Originally, values were derived from data for the U.S. Midwest; however, because that data cannot be assumed to be representative elsewhere, the equation has to be adapted to the sites where it shall be applied. It is quite data demanding to fit the equation to local conditions; therefore, simpler empirical models, such as the Soil Loss Estimation Model for Southern Africa (SLEMSA), were developed. The USLE was developed further into the Revised Universal Soil Loss Equation (RUSLE)[49] (Renard et al. 1991). The Water Erosion Prediction Model (WEPP) allows for more complexity (Laflen, Lane, and Foster 1991) but is even more data demanding than RUSLE and USLE.

Other possible approaches assess the impact of erosion experimentally on fields or in a laboratory, and some studies used assessments of soil erosion based on local expertise and subjective assessments (see, for example, Alfsen et al. 1996; McKenzie 1994).

Most studies have indicated a positive effect of conservation measures on farm income or profitability. Byiringiro and Reardon (1996) found that an increase in soil conservation investment per hectare from low to high increases the marginal value product of land by 21 percent. Kaliba and Rabele (2004) analyzed a positive impact of conservation measures and concluded that farmers

47 Other models are Crop Environment Resource Synthesis (CERES); Agricultural Production Systems Simulator (APSIM); Soil Conservation in Agricultural Regions (SCAR); and Productivity, Erosion, and Runoff Functions to Evaluate Conservation Techniques (PERFECT) as reviewed in Enters (1998).

48 The relationship between topsoil depth and crop yield can be estimated using the Mitscherlich-Spillman production function, exponential functional forms (Lal 1981), and various other functional forms (Ehui 1990; Walker 1982; Taylor & Young 1985; Pagiola 1996; Bishop & Allen 1989).

49 Recent application of RUSLE has been done by Pender et al. 2006; Nkonya et al. 2008a).

gain more from soil conservation measures than from using inorganic fertilizer alone. Other studies have found positive impacts of conservation under certain conditions, such as plot size and slope (Adegbidi, Gandonou, and Oostendorp 2004), rainfall conditions (Bekele and Drake 2003), type of conservation measure (Kassie et al. 2008; Bravo-Ureta et al. 2006), and a reduced variability of yields (Kassie et al. 2008). All of these studies analyzed the on-site effects of land-degrading and land-conserving measures. Because degradation problems tend to be site specific and the adoption of soil conservation measures depends on the decisions of individual farmers, most case studies are applied at the farm level (Lutz, Pagiola, and Reiche 1994). To assess the extent of degradation, most studies limit themselves in their analysis to the impact of certain processes of land degradation on agricultural yields, such as soil erosion rates,[50] nutrient depletion,[51] soil compaction, or salinization.

Adoption Models

Adoption of sustainable land management techniques and investments in soil conservation practices depends not only on monetary profits but also on factors that affect a more general definition of *benefits*. Formal analysis conducted by McConnell (1983) shows how it may be optimal for farmers to make production choices in which rates of soil depletion exceed what would be socially optimal. Inefficiencies in capital markets, for instance, may truncate farmers' planning horizons, thus introducing soil-depleting biases, or it may affect farmers' rates of time discount, so that it exceeds the market rate of return on capital. Furthermore, as De Pinto, Maghalaes, and Ringler (2010) pointed out, the combination of inefficient markets and farmers' risk attitudes can lead to a low rate of adoption of sustainable land management practices. The growing body of empirical work uses socioeconomic characteristics of farmers (age, gender, education, and so on), land characteristics and natural conditions (farm size, plot size, slope), farm management practices, and institutional aspects (support programs, access to credits, and so on) to analyze farmers' choices.[52] A comprehensive review of the institutional factors and policies that influence land use decisions are given in a specific section of the report. This section reviews empirical findings of many studies on the adoption of land conservation

---

50 Measured as loss of soil (in tons per hectare per year).

51 Often assessed as nutrient balances (see, for example, Stoorvogel et al. 1993; Stocking 1986; Smaling et al. 1996; Craswell et al. 2004).

52 The decision of whether a farmer applies conservation can be modeled with discrete choice models, such as probit (Amsalu & de Graaf 2007), tobit (Pender & Kerr 1998), or logit (Place & Hazell 1993). The assumption underlying these types of models is that an economic agent, often the household or the farmer, chooses inputs and technology with the goal of maximizing utility or profit..

measures and provides insight into the variety of factors that have been found to significantly influence land use decisions.

Adoption models are used to explain the sources of variability in adopting soil conservation measures, as some farmers adopt conservation measures while others do not even though doing so may be profitable to them. In their study on the impact of soil conservation on farm income, Bravo-Ureta et al. (2006) summarized many studies on factors determining adoption decisions. Concerning farmer and household characteristics, awareness of the erosion problem was found to be a significant factor by various authors (Norris and Batie 1987; Hopkins, Southgate, and Gonzalez-Vega 1999; Shiferaw and Holden 1998; Pender and Kerr 1998), as was the perception of the profitability of conservation measures (Amsalu and de Graaf 2007). Results on age show mixed influences. Amsalu and Graaf (2007) found that older farmers are more likely to adopt conservation measures, whereas Norris and Batie (1987) identified younger farmers. Other authors did not find significant effects of age on adoption (see, for example, Bekele and Drake 2003). In addition, results regarding farmer's education are inconclusive. For example, Pender and Kerr (1998) and Tenge et al. (2004) identified a positive influence of education on investments in indigenous conservation measures. Farm characteristics, farm size, plot size, slope, and location of plots are important factors that affect the adoption of soil conservation. Farm size also has a positive effect according to studies by Norris and Batie (1987), Pender (1992), Bravo-Ureta et al. (2006), Amsalu and de Graaf (2007), and others; however, other authors found that farm size has insignificant or negative effects. Plot size is expected to influence conservation positively, because conservation structure will need a larger proportion of the plot and will thus reduce the area under production, which may not be enough to compensate for the area lost when plots are small (Bekele and Drake 2003). Slopes have mostly a significantly positive effect on conservation, as found in Winters et al. (2004), Bekele and Drake (2003), Shiferaw and Holden (1998), Amsalu and de Graaf (2007), and Nyangena and Köhlin (2008).

Many institutional- and policy-related factors have been analyzed so far. The study by Lutz, Pagiola, and Reiche (1994) found that landownership positively influences the adoption decision. Pender and Kerr (1998) showed negative effects of sales restrictions of land and tenancy on conservation investments. Empirical studies show that some farmers without legal titles invested in soil conservation, while others with legal titles may have not (Lapar and Pandey 1999). The existence of well-defined and enforceable property rights to land seems to be a necessary but not sufficient condition to adopt soil conservation technologies (Anlay, Bogale, and Haile-Gabriel 2007). Regarding land rights, Place and Hazell (1993) tested in Ghana, Kenya, and Rwanda whether the indigenous land rights systems are a constraint on agricultural productivity and found that land rights do not significantly influence land improvements—with a few exceptions depending on the region analyzed. Other factor market imperfections, such as access to credit, show mixed results on adoption (Napier 1991; Pender 1992). Hopkins, Southgate, and Gonzalez-Vega (1999) and Pender and Kerr (1998) identified a positive effect of off-farm income on adoption, whereas Amsalu and de Graaf (2007) found a significantly negative impact in their study on the continued use of stone terraces. Because labor availability is important, especially for labor-intensive conservation measures, membership in a local organization with labor-exchange arrangements leads to higher adoption (Lapar and Pandey 1999). The presence of agricultural extension services increases the availability of information on land degradation and conservation measures and, hence, also increases the likelihood of farmers to adopt conservation (Bekele and Drake 2003).[53]

An interesting feature of the study by Amsalu and de Graaf (2007) is that they analyzed adoption as well as the continued use of soil conservation measures; they found that factors influencing adoption and continued use of the stone terraces in the Ethiopian highlands are not the same. Actually, long-term adoption is more relevant to analyze, as it allows maintaining soil productivity in the long run. Further work on long-term adoption and use is required.

Estimation of the Farm-Level Costs and Benefits of Land Degradation

The formal analysis of land degradation began in the 1980s with optimal control models (developed by Burt [1981] and McConnell [1983]) that modeled the decision of individual farmers. These models aimed at maximizing the net present value of the agricultural output in order to find the optimal rate of land degradation. Optimization models have various advantages over partial budgeting (as described earlier), because they are not limited to a fixed number of alternative options; rather, they allow for more flexibility in adaptation. Furthermore, they help select the optimal profit that maximizes land management practice. An important advantage of optimization models[54] is that they deliver shadow values, representing changes in profits, which are derived from marginal changes in resources (Mullen 2001). De Graaf (1996) referred to these models as investment models, because they focus on profit maximization as the sole objective of a farmer. This may be the case for large commercial farmers, but peasants and farm households may behave differently (Kruseman et al. 1997). Households may decide on farming activities and soil conservation

53 Other factors that may have an impact on adoption that were analyzed in studies are the influence of family labor, livestock, soil fertility, soil depth, erosion status, agricultural inputs, management practices, and diversification, as well as other physical, personal, economic, or institutional factors.

54 These initial models were further developed by various authors, such as Barbier (1990), Barrett (1991), Miranda (1992), Clarke (1992), and Coxhead (1996).

measures in a way that maximizes utility (which includes risk considerations and accounts for a subsistence level of income) rather than profits. These kinds of models appear to be more appropriate when farm households and farmers producing mainly for subsistence or at a smaller scale need to be considered.

In summary, the models described here are flexible and allow the integration of economic and biophysical conditions and feedbacks into the local economy; thus, they explicitly model the relationship between production and degradation.[55] These models can be designed for multiple periods (see, for example, Shiferaw and Holden 2005) or for a single year (see, for example, Day, Hughes, and Butcher 1992). Furthermore, they incorporate the impacts of various market imperfections (see, for example, Shiferaw and Holden 2001; Holden, Shiferaw, and Pender 2004) and can incorporate the impacts of policies and subsidies (for example, access to credits in Börner 2006), as well as combined impacts of land degradation, population growth, and market imperfections (Holden and Shiferaw 2004). These advantages come at a cost: The models are data demanding and highly complex. Therefore, many studies have evolved that use the cost–benefit analysis framework as presented earlier.

Many farm household models have been developed to address land conservation decisions in Ethiopia. Mengistu (2011) developed a farm household model to analyze selected policy incentives and technology interventions on land quality and income of small farm households in the Anjeni area of northwestern Ethiopia. Another farm household model for Ethiopia's Ada district was developed by Shiferaw and Holden (2000, 1999, 1998). Major model activities include crop production on uplands and lowlands with three levels of fertilizer use, with and without conservation measures; crop sale and consumption; seasonal family labor; labor hiring; leisure; and livestock production and activities to account for future negative impacts of soil erosion. A more recent application developed by Holden, Shiferaw, and Pender (2004) incorporated access to nonfarm income and analyzed its impact on soil-conserving behavior.

Estimation of Off-Site Costs and Benefits

Off-site costs are related to the effects of land degradation on the surrounding environment, in particular downstream impacts of land degradation. However, they also comprise global effects on services, such as carbon sequestration, biodiversity, and food security. We have already described in detail the possible off-site effects arising from land degradation.

Many studies were conducted to come up with cost estimates for the impact of sedimentation caused by upstream soil erosion on agricultural land.

55 Based on earlier versions developed by Shiferaw et al. (2001, 2002) and Shiferaw (2004), and Holden, Shiferaw, and Pender (2002).

Lost soil drains into major dams and reservoir systems that provide irrigation, hydroelectricity, or flood control services. Siltation in reservoirs reduces water storage and electricity production, shortens the life span of dams, and increases their maintenance costs. Heavy sedimentation frequently leads to river and lake flooding. As a first step, it is important to quantify the amount of sedimentation caused by land degradation, particularly by agricultural soil loss. Unfortunately, other activities, such as mining, construction works, or unpaved roads, also contribute to sedimentation, which means the relative impact of its causes is difficult to determine. The dynamics of sedimentation processes are complex: Sediments may be temporary or permanently stored along the waterways, and it takes time for impacts from increased sedimentation in reservoirs to become visible. Furthermore, a strict categorization of sedimentation as a cost factor may not be adequate, because sedimentation may be beneficial to downstream users by providing farmers with fertile, nutrient-rich soil (Clark 1996) or by serving as construction input (Enters 1998).

Existing studies have employed a range of methods for the valuation of off-site effects of land degradation—chiefly, erosion. The following review sorts studies according to the kind of damage or off-site cost that is valued. Appendix 3 provides a brief, systematic overview of relevant studies estimating off-site costs in a table format.

Damage on Reservoirs for Irrigation and Hydropower

Various estimations of the costs due to the sedimentation of reservoirs were conducted by Cruz, Francisco, and Tapawan-Conway (1988) in the Philippines and Magrath and Arens (1989) in Java, Indonesia. Wiggins and Palma (1980), as reviewed in Clark (1994), estimated the impact of reservoir sedimentation on hydropower generation. The loss in generation capacity is valued in terms of the least-costly alternative source of power, which is electricity generated by thermal power stations (Clark 1994). Abelson (1979), as reviewed in Clark (1994), analyzed the impact of sedimentation on irrigation water by estimating the decline in the output of dairy farms that use water for irrigation. The value of water lost due to lower storage capacity was then calculated from the social value of milk production based on world market prices (Clark 1994). Vieth, Gunatilake, and Cox (2001) estimated the off-site costs of soil erosion in the Upper Mahaweli watershed in Sri Lanka. The reduced capacity of the reservoir to store water for irrigation is valued by the reduction in irrigated area, the impact on hydropower production, and the increased water purification costs. Hansen and Hellerstein (2007) valued the impacts of soil conservation on reservoir services in terms of reduced dredging costs for a one-ton reduction in erosion across the 2,111 U.S. watersheds.

Navigation Damage

Sedimentation due to soil erosion may also have negative impacts on navigation in waterways. Gregerson et al. (1987) analyzed this impact for the Panama Canal. The cost of sedimentation is valued by the cheapest alternative method to deepen the canal using dredgers. Hansen et al. (2002) quantified the costs of soil erosion to downstream navigation using a damage-function approach.

Water Treatment

As mentioned in the study of Vieth, Gunatilake, and Cox (2001), sedimentation causes a higher level of turbidity, which increases the cost of water purification. Moore and McCarl (1987) used the cost of extra chemicals that are needed to coagulate the particles in the water to value this off-site impact. Holmes (1988) used a hedonic cost function to estimate the cost of water purification. Nkonya et al. (2008b) included the increased costs of production of clean water due to soil.

Flooding and Aquifer Recharge

Richards (1997) estimated off-site benefits related to conservation practices in the Tequila watershed in Bolivia. The main off-site benefits were identified as flood prevention and increased infiltration of water in the soil (due to reduced runoff) and thus higher water availability in the aquifer.

Recreational Damage Estimates

Soil erosion can also have recreational impacts, as particles and pollutants reduce the water quality and freshwater fishing possibilities. In addition, siltation and weed growth interfere with boating and swimming activities, thus decreasing the site's recreational value. Clark (1985) indicated recreational damages on freshwater fishing, marine fishing, boating, swimming, waterfowl hunting, and accidents. Using travel cost models, Feather, Hellerstein, and Hansen (1999) estimated the benefits of fresh-water-based recreation, wildlife viewing, and hunting of the Conservation Reserve Program in the United States. Bejranonda, Hitzhusen, and Hite (1999) examined property values at 15 Ohio state park lakes to analyze the effect of sedimentation and found higher property values on lakes with less sedimentation.

Comprehensive Studies

Comprehensive studies, including valuation of several off-site costs at once, were carried out by a number of authors. An early study by Clark et al. (1985) and a study by Pimentel (1995) both calculated various off-site costs associated with wind and water erosion in the United States. Tegtmeier and Duffy (2004)—based on previous work by Clark (1985) and Ribaudo (1986), among others—

provided national estimates of total annual cost damages attributable to water-based soil erosion in the United States.

Pretty et al. (2000) provided an assessment of the total external costs of agriculture in the United Kingdom. Krausse et al. (2001) estimated the economic costs of sedimentation effects in New Zealand, and and Hajkowicz and Young (2002) did the same in Australia between 2000 and 2020. More recently, nonmarket valuation approaches, such as contingent valuation method and choice experiments, have been used to value several off-site effects. A study by Colombo, Calatrava-Requena, and Hanley (2003) used contingent valuation and found that a majority of the catchment's population is willing to pay to reduce off-site damages. Colombo, Hanley, and Calatrava-Requena (2005) also conducted a choice experiment in the Alto Genil and Guadajoz watersheds in southern Spain. Respondents were found to care about the negative effects of soil erosion on surface and groundwater quality, landscape desertification, and flora and fauna. Social impacts (rural employment) also turned out to be important. A positive willingness to pay could also be found regarding the size of area benefiting from soil erosion control programs.

Global Benefits

Nkonya et al. (2008b) also indicated global off-site benefits associated with conservation measures that increase the biomass on the field and, hence, that lead to increased carbon sequestration. Carbon accumulation due to conservation measures is estimated at 0.2 to 0.7 tons of carbon per hectare per year (Vagen 2005). Earlier studies of carbon sequestration revealed a value of $3.50 per ton of carbon stored, though this value is debatable and is bound to fluctuate according to the evolution of carbon markets.

Global Off-Site Costs

Basson (2010) estimated the annual global cost of siltation of water reservoirs is about $18.5 billion for storage structures, with the replacement costs of silted-up reservoirs accounting for a little more than 50 percent of the total cost (Figure 3.7). The annual loss of hydroelectric power (HEP) and damage to HEP infrastructure is about $5 billion; the loss due to the reduction of irrigation reservoir capacity is about $3.5 billion. These losses do not include other losses due to siltation, such as the loss of potable water and related health effects. Thus, the estimated losses could be regarded as being conservative.

Figure 3.7—Global loss due to siltation of water reservoirs (US$billion)

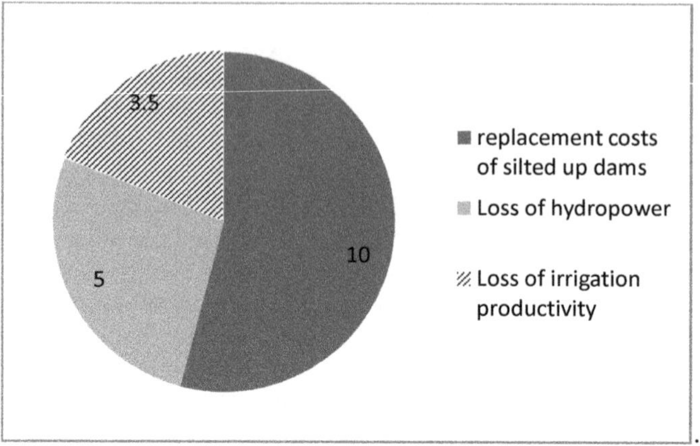

Source: Basson 2010.

Estimation of the Indirect Costs of Land Degradation

The indirect costs of LD represent their impacts across all sectors of the economy—for instance, through price transmission mechanisms or transactions on the input markets—as well as their human impacts (migration, food security, poverty, and so on). Thus, economywide effects of soil erosion do not simply equal the lost production multiplied by a given price. Due to the linkages across the sectors of the economy—as well as between supply and demand—the production, demand, prices, and trade of all commodities, beyond the crops directly affected by soil loss, will also be affected. Further, in several countries, agriculture is the main component of rural households' livelihood strategies; thus, another indirect effect of land degradation is its impact on poverty and poverty rates—for instance, through income effects. Several links among the environment, agriculture, and poverty are reviewed in Vosti and Reardon (1997). Also, lower production levels of particularly staple foodcrops can cause or increase food insecurity problems, especially when population is growing (Diao and Sarpong 2007).

Indirect effects are difficult to assess; therefore, only a few studies have attempted to measure at least some of these costs. Some of those studies are briefly reviewed here. Alfsen et al. (1996) estimated the impact of soil erosion on GDP, imports, exports, and consumption using a computable general equilibrium (CGE) model of the Nicaraguan economy to explicitly consider interlinkages between agricultural activities and other parts of the economy. Model results suggest that there are significant production impacts due to soil erosion and that this also affects trade, labor, private consumption, and investment. Gretton and Salma (1997) estimated an econometric model of the impact of degradation (irrigation salinity, dryland salinity, soil structure decline, and induced soil acidity) on the agricultural output and profitability, based on state-level data from New South Wales, Australia. The econometric analysis indicated that agricultural output and profits depend on the type of degradation. The results suggest that the expansion of some farming systems, and the associated increased degradation due to salinity, provides a net increase in production and profits in the medium term, whereas soil structure decline and induced soil fertility lead to negative net effects.[56] Other studies using a larger-scale approach include village CGE (Holden and Lofgren 2005), national CGE (Alfsen et al. 1996), and multisector CGE (Coxhead 1996).

Effects of soil loss on the economy and on poverty were estimated by Diao and Sarpong (2007) in Ghana. They estimated that the declines in the national and rural poverty rates between 2006 and 2015 were 5.4 and 7.1 percentage points less, respectively, when soil loss is taken into account. Furthermore, the projected slowdown in production growth of staple foodcrops, such as maize, would cause food security problems, given that the population of Ghana is expected to grow at 2 percent per year (Diao and Sarpong 2007). Sonneveld (2002) assessed the effects on food security in Ethiopia based on scenarios: Food availability per capita dropped from 1971 kilocalories per day in 2000 to 685 kilocalories per day in 2003 due to water erosion, assuming no additional conservation activities (Scenario 1). To assess the impact of LD on migration and the resulting costs, CSFD (2006) proposed a methodology for its valuation. They differentiated between direct costs and indirect costs as a function of the places of origin and arrival and accounted for the costs and benefits (in receiving country) of migration in order to give a complete inventory of the costs and benefits associated with migration. However, they did not calculate the actual costs of land-related migration.

3.3 Summary

The literature review shows a range of different methodologies that have been applied in the past to study the costs of land degradation and conservation. Table 3.2 summarizes which kinds of methods can be used to assess different types of costs of LD. Although more possibilities for valuing costs of land degradation exist, the ones listed here are the most frequently used and were reviewed before. Bringing together the different costs and value types to fully assess total costs and benefits over time and their interactions can be done within the framework of a cost–benefit analysis and mathematical modeling. How to match the concept of ecosystem services to their economic valuation is

56 Other structural econometric models were developed for example by Pender and Gebremedhin (2006).

an issue that requires further research. There is also a problem of double-counting, which needs close attention, as ecosystem services are not easily split up into particular benefits that can be valued and then aggregated.

Table 3.2—Economic valuation techniques for the estimation of various types of costs

Cost type or type of economic value	Economic valuation technique
On-site, direct cost, use value	Productivity change approach, replacement cost approach
Off-site costs	Damage costs, avoidance/mitigation costs, stated preferences techniques, travel cost method
Nonuse values, existence value	Stated preferences techniques, hedonic pricing, travel cost method
Indirect costs	Mathematical modeling, econometric approaches

Source: Compiled by authors.

Studies that aim to estimate the costs of land degradation have used methods and approaches that vary widely in their underlying assumptions, and their results are thus difficult to directly compare. Bojö (1996) identified 10 different dimensions of costs based on an extensive review of studies on land degradation conducted in Sub-Saharan Africa. The author pointed out how the definition of costs and how the measurement vary across studies, as well as how the scale of analysis changes across studies, ranging from plot-level soil analyses to watershed or catchment areas and to production losses computed at the national and global level. Few studies have examined the variety of off-site costs of land degradation, in addition to on-site costs of land degradation (Nkonya et al. 2008b, Magrath and Arens 1989); rarely are indirect costs considered (Requier-Desjardins, Adhikari, and Sperlich 2010). Even macrolevel studies seldom include degradation-related issues, such as poverty, rural living standards, and poor education (Requier-Desjardins 2006), or the costs associated with the impacts of government policy and trade agreements. Bioeconomic modeling (Barbier 1996; Shiferaw and Holden 2005) and farm household models have great appeal because they allow linkages to rural livelihoods, governmental policies, price distortions, and so on. However, they turned out to be highly complex and data intensive. Therefore cost–benefit analysis is often the preferred approach (Enters 1998). In general, a lack of data on the level of degradation and conservation prevails, because many of the key processes are difficult to measure in terms of individual components and interactions between them over time and across space (Berry, Olson, and Campbell 2003). A quantification of the resulting productivity losses, as well as off-site effects, is already challenging and would be required over time, taking into account dynamics, aggregation over time, and nonlinearities.

Investments to mitigate degradation may not always be profitable at the farm level, especially in areas already highly degraded or with fragile land, where crop production is not low. In addition, poor farmers have much higher discount rates, so that conservation measures tend to be nonprofitable for them. However, off-site effects are not relevant from a farmer's perspective, and carrying capacity loss is omitted (Requier-Desjardins 2006).[57]

There is a need to define a comprehensive framework for an assessment that would include consideration of environmental, social, institutional, and economic factors (Requier-Desjardins, Adhikari, and Sperlich 2010). Such a framework would require a common definition of all relevant costs and would need to cover the total economic value of land resources in order to accommodate the full range of impacts of LD on terrestrial ecosystem services on and off site. The analysis also needs to be comprehensive and should include social impacts, such as migration, impacts on poverty, and rural livelihoods, as well as economic linkages with other sectors. It also must capture impacts of institutions, economic and social policies, and so on. The indirect costs of LD arising from such linkages are currently poorly understood.

57 Cohen, Brown, and Shephard (2006) suggested energy synthesis, a framework that integrates all flows within a system of coupled economic and environmental work in biophysical units (embodied solar energy/solar emjoules). Hein (2002) provided a framework for the analysis of the costs of land degradation based on the environmental function approach by de Groot (1992) and de Groot, Wilson, and Bouman (2002).

4. Actors, Incentives, and Institutions Governing Land Management

There are ecological (naturally occurring)[58] and anthropogenic (human-induced) causes of land degradation (LD) (Mainguet and Da Silva 1998).[59] The latter is governed by the corresponding "rules of the game" (institutions), which act as constraints and incentives for actors' decisions (North 1990). Understanding both the relevant institutions and the actors' incentives is therefore crucial to address the causes of LD efficiently and sustainably.

Institutions play a mediating role between the drivers of land degradation and those of land improvement. As shown in Figure 4.1, two different outcomes could occur under the same set of drivers of land degradation. For example, studies have shown opposing impacts of population density and land degradation. Whereas Grepperud (1996) and Cleaver and Schreiber (1995) found a positive relationship between population density and land degradation, Bai et al. (2008b) and Tiffen, Mortimore, and Gichuki (1994) found a positive relationship between population density and land management in the world and in Kenya, respectively. Boserup (1981) suggested that population pressure on the land will stimulate the development and adoption of new production technologies. Population growth thus provokes technological change, resulting in higher productivity of agricultural workers; in turn, this creates a positive feedback from population growth and increased density to agricultural development. Von Braun et al. (1991) found that in a very densely populated area in northwestern Rwanda with a high population growth rate of 4.2 percent, substantial indigenous mechanisms for increasing labor productivity under increased land scarcity existed. A 10 percent increase in the person–land ratio resulted in only a 3.6 percent decline in labor productivity, showing that labor productivity declines less than proportionally as farm size decreases. However, even a high rate of technological change in agriculture could not fully compensate for the area's high population growth. Further, the success of land management programs depends on how institutions are taken into account when planning land management programs. For example, empirical evidence has shown severe depletion of protected national forest reserves, which thus excluded local institutions and communities from participating in the management and sharing of the forest's benefits.

58 An example of an ecological factor of land degradation is rolling topography, which causes soil erosion.

59 Several authors highlighted that it is mainly the interaction of the land with its users that leads to any kind of land degradation resulting in serious social problems (Vlek, Le, & Tamene 2008; Johnson & Lewis 2007; Blaikie & Brookfield 1987; Spooner 1987; Barrow 1991). See also Section 2 for a discussion of causal links.

Figure 4.1—The mediating role of institutions

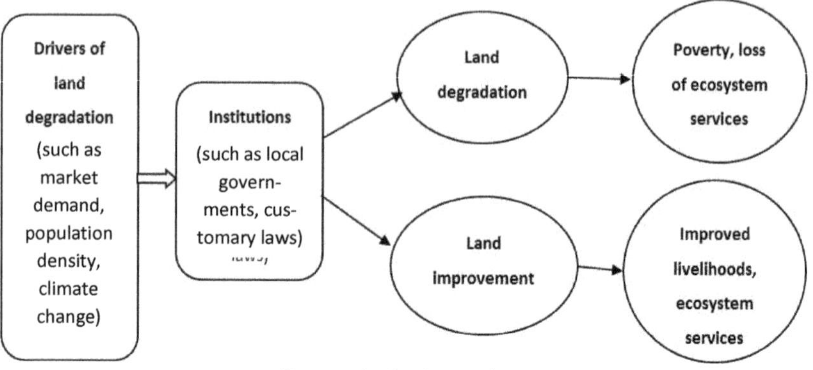

Source: Author's creation.

A study by Ostrom and Nagendra (2006) showed better management of forest resources by communities and their institutions than the protected forests managed by central governments.

This section discusses the most important actors and their respective incentives, as well as major institutions and policies affecting land management and productivity. This section comprises a review of the poverty–land management relationship, as well as scrutiny of the causalities between secure property rights and land management decisions. This discussion shall deepen the understanding of the factors that affect land management decisions. Finally, a framework of analysis for social and economic causes of LD will be presented to help identify and address the key causes of nonaction in response to LD.

4.1 Roles and Levels of Institutions Governing Behavior

Institutions, defined here as "the formal and informal rules governing economic production and exchange" (North, 1991), serve to eliminate conflicts of interest or ambiguity by defining what it is that people can expect from others (Colson, 1974, in North, 1991).[60] Institutions define the structure within which (economic) actors make decisions. Within different life situations, individuals face a complex combination of institutions at different levels, and these institutions help shape their choices for action and exchange. One way of conceptualizing this is to distinguish institutions according to their persistence

60 In this context, it is important to distinguish between this institutional economic definition and the common use of the word institution, as it is frequently used when talking about financial institutions, international institutions, and so on. The latter are organizations that often can create, shape, or influence institutions—for example, by issuing laws, contracts, or conditions for exchange.

and velocity of change.[61] Williamson (2000) distinguished four levels of institutional analysis: On the top lies the *social embeddedness* level, where the norms, customs, mores, traditions, and so on are located. This level is taken as a given by most institutional economists, because it takes extremely long to change and such change is a collective, social process rather than being orchestrated by a group of actors. The second level is referred to as the institutional level, where structures are partly the product of evolutionary processes and partly designed. In this context, Williamson (2000) coined the following phrase: "The opportunity for first-order economizing: get the formal rules of the game right" (598). Influenced by both old and recent history, this layer includes the judicative, legislative, and executive structures of government, as well as their share of power (for example, the degree of federalism). The structure and enforcement of property rights and contractual arrangements are another important part of this level.[62] Next is the third level, where the institutions of *governance* are located. At this level, the focus shifts to the contractual arrangements between interacting parties—that is, to the organization of economic transaction. As contracts or equipment become renewed, such transactions between governance structures (for example, firms) become periodically reorganized—on the order of a year to a decade. This is referred to as "the second-order economizing: get the governance structures right" (Williamson 2000, 599).

These three levels of discrete analysis of governance structures must be distinguished from the fourth layer, which is the level where *neoclassical economic analysis* and agency theory work to determine resource allocation (wages, employment, prices, and so on). At this level, firms are typically depicted as production functions and analyzed as an "optimality apparatus," using marginal analysis, wherein adjustments to prices and output are more or less continuous. This lowest level is embedded in higher levels of institutions. The four levels of analysis are illustrated in Figure 4.2.

61 Such distinction is also of help for policy design, as it might indicate where potential quick fixes can be found, as well as where longer-term difficulties are to be expected. The final part of this section returns to this idea.

62 These first-order choices are, without doubt, important for the outcome of an economy (Coase 1992 and Olson 1996, cited in Williamson, 2000). Still, cumulative change of such structures is very difficult to orchestrate, though it occasionally takes place if historic events lead to a sharp break from established procedures (disasters, wars, crises, and so on.).

Figure 4.2—Economics of institutions

	Level	Frequency (years)	Purpose
L1 social theory	Embeddedness: informal institutions, customs, traditions, norms, religion	100-1000	Often noncalculative; spontaneous
L2 economics of property rights/ positive political theory	Institutional environment: formal rules of the game– especially property (polity, judiciary, bureaucracy)	10-100	Get the institutional environment right. 1st order economizing
L3 transaction cost economics	Governance: play of the game – especially contract (aligning governance structures with transactions)	1-10	Get the governance structures right. 2nd order economizing
L4 neoclassical economics/ agency theory	Resource allocation and employment (prices and quantities; incentive alignment)	continuous	Get the marginal conditions right. 3rd order economizing

Source: Williamson 2000, 597.

4.2 Type of Actors Involved in Land Management

As discussed earlier, LD is attributed to a combination of anthropogenic and natural processes, both of which are guided by complex interactions. Anthropogenic processes are characterized by complexity and two-way causalities. Dryland livelihoods, for example, have been based on a flexible combination of hunting, gathering, cropping, animal husbandry, and fishing (multiaction), whereas livelihood strategies change in time and space to adapt to new economic possibilities, in response to environmental or climatic changes, or as a result of war- or drought-induced migration (Berry 1993; Niemeijer 1999; Robbins 1984). Land use changes are thus both responses to changes in the provision of ecosystem services and drivers of changes in this provision (Safriel and Adeel 2005).

Having revised a number of global assessment models for environmental change, Leemans and Kleidon (2002, p. 215) concluded "that the next generation of models can only accomplish a comprehensive assessment of desertification through a simultaneous consideration of both the socioeconomic and biophysical dimensions." Although developing such a comprehensive assessment of the theoretical framework of institutional economic models is beyond the scope of this report, accepting the complexity while conceptualizing these processes remains its purpose.

To better understand the anthropogenic or human-induced LD processes, the matrix in Table 4.1 depicts categories of main actors. The first column lists ten groups of actors. The second column suggests criteria for typologies, which helps conceptualize these groups. Such typologies can be useful in looking at differences in incentives, power, or (expected) behavior, which in turn helps understand or even predict land use (for example, after policy or price change). The third and fourth columns give examples of the actors and their respective use or impact on land use. We purposely did not attempt to include the major incentives of each group of actors, as that degree of detail does not seem necessary given the complex interaction of social, economic, and environmental dynamics governing land use patterns (Leemans and Kleidon 2002; Herrmann and Hutchinson 2005; Lambin 2002). Although such a list is not exhaustive, it can help group the actors involved and provide a better understanding of their rationale.

Table 4.1—List of actors affecting land use decisions

Actors	Potential criteria for typology	Example	Type of use/impact on use
(Direct) Land user	Individual — collective, private — public, commercial — subsistence	Farmer, forester, herder	Agriculture, livestock raising, forestry, tourism, industrial use
Landowner	Individual — collective; private — public	Community, landlord, farmer	Renting out land, selling land, monitoring of user
Government	Local — national	Ministries, municipalities	Design land policies, environmental laws; market access; research and development for agricultural uses; subsidies for inputs
Customary institutions	Local — collective	Village or clan chief:	Customary land tenure administration
Industries	National — international; different scales and sectors	Commercial agriculture producers; industries with negative externalities on land and water	Large-scale production of food or cash crops; leather production and tannery; paper industry
Beneficiaries of ecosystem services	Local, national, and international; legitimate — illegitimate; public — private; payment	Tourists; breweries; consumer of drinking water; herders	Recreation; water consumption

Table 4.1—Continued

Actors	Potential criteria for typology	Example	Type of use/impact on use
NGOs	National — international; direct — indirect involvement; area of operation; budget size	Wildlife and biodiversity conservation; poverty alleviation and rural development; off-farm employment and education; legal advice	Direct land use (for example, buying of rain forest); impact on land use (for example, the Forest Stewardship Council); training and extension services
International organizations and development agencies	National — international; different scales, budget, and sectors; mandate	UNCCD; WB; GTZ; KFW; DFID/UK Aid; FAO, ADB, AfDB	Financial and technical assistance to direct users; advice to governments and policymakers; legal enforcement
Research institutes and academia	National — international; different size and focus	CGIAR; Universities; IFPRI; ZEF; DIE	Research on and monitoring/measurement of LD; understanding drivers and informing political decisionmaking

Source: Author's compilation.

4.3 Institutions for Land Management: What Matters, Why, and How to Improve?

Based on aforementioned layers of institutions and groups of actors, and building on earlier discussions of the causes of LD (Section 3), this section will discuss (i) which incentive structures matter for actors, (ii) what policies are most affecting LD, and (iii) how to approach institutional design for better outcomes.

Box 4.3—Customary institutions matter: An example from the Maasai in East Africa and Buddhists in Burma

The Maasai of Kenya and Tanzania

The Maasai have a unique, environmentally friendly custom that sets them apart from surrounding communities in East Africa. The Maasai are pastoral communities with strong traditional livelihoods, and they have outlived the onslaught of modernity. One of the strong features of the Maasai tradition is that they do not eat wild game meat (Asiema and Situma 1994) or cut a live tree. This shows their strong environmentally friendly tradition, which other surrounding communities do not have. Before the colonial period, the Maasai lived in what are now game parks and harmoniously shared the ecosystem services with wildlife. The Maasai regard trees as landmarks of water sources, cattle routes, and medicinal herbs (Ole-Lengisugi 1998). This is one of the reasons that the government of Tanzania allows only the Maasai to live in the game parks.

Box 4.4—Continued

The Green Monks of Burma

One of the Buddhists' key tenets of environmental friendliness is their compassion toward all living beings. This tenet, called *metta*, states that all Buddhists should abstain from destroying any living being (Nardi 2005). Due to this, Burma (Myanmar) is one of the countries in Asia and Oceania that have largest forested area; the others are China, Australia, Indonesia, and India (FAO 2011). However, recent economic hardships have led to deforestation in Burma. From 1990 to 2000, the deforestation rate was 1.2 percent; this rate fell to 0.9 percent during 2000–2010, which is more than twice the deforestation rate of 4 percent in Southeast Asia (FAO 2011).

How to Improve Institutional Design to Lower LD

Having discussed different areas of policy affecting LD, which policy and institutional design is most appropriate for enhancing the role of institutions to address LD?

Decentralization, Involvement of Local Communities, and Capacity Building

One of the major reasons behind the failure of centralized governments to effectively manage land resources is the lack of involvement of local communities in managing and benefiting from natural resources and the financial resources for managing resources (Gibson, Williams, and Ostrom 2005). This exclusion creates alienation, which in turn leads to poor cooperation between local communities and natural resource managers. Therefore, the aim is that local people should participate jointly in problem identification and in the design of culturally appropriate and sustainable solutions. Participatory approaches usually imply active engagement of local people and agencies that goes beyond eliciting the views of individuals, extending to processes of interactive dialogue, collective learning, and joint action. This type of approach values local knowledge in addition to the usual scientific and technical knowledge. A participatory approach may help deal with the complexity of land management decisions through the use of more creative tools and techniques, rather than through centralized management, which tends to have limited local knowledge and, in developing countries, limited human and financial resources to enforce natural resource management regulations.

In some instances, decentralization efforts took center stage in efforts to address the poor management of natural resources by central governments take place (Agrawal and Ribot 1999; Devas and Grant 2003), which is a recognition of the role of local communities and their institutions. A long-term study that examined the effectiveness of various forms of institutional settings to manage forest resource, conducted by the International Forestry Resources and Institutions (IFRI), generally observed that locally managed forest resources are better managed than centrally managed forest resources (Ostrom and Nagendra 2006). It is estimated that about one-quarter of forests in developing countries are under some form of community-based forest management (FAO 2011; CIFOR 2008). The share of community-based managed forests is also increasing due to decentralization efforts and promotion of community forest management by nongovernmental organizations (NGOs) and international organizations (FAO 2011).

As expected, however, performance of local communities in effectively managing natural resources depends on a variety of factors, and poor human capacity remains a key challenge. Hence, one condition for successful community resource management is organizational supply, which is determined by the presence of community members or organizations with substantial leadership or other assets (Ostrom 1990). A study in Uganda, for example, showed that communities with government programs or NGOs dealing with agriculture and the environment in communities had a higher propensity to enact bylaws on natural resource management (Nkonya, Pender and Kato 2008) The same study observed significantly higher compliance with laws and regulations enacted by local councils, as compared with those enacted by legislative bodies higher than the local council (Nkonya, Pender and Kato 2008). These findings further demonstrate the importance of local-level organizations in reducing enforcement costs and increasing sustainability. Nonetheless, land degradation cannot be changed by participatory approaches and community action alone; public investment in infrastructure and other policies that support land management are also needed (Koning and Smaling 2005).

Government policies and institutions also play a big role in determining the effectiveness of local community institutions. The major policies have been discussed earlier. In this section, we show the role those policies play in fostering local institutions. Policies that foster and build capacity of local government generally enhance the effectiveness of local institutions. Decentralization, in particular, is key to achieving this goal. A study covering four African countries compared the number of bylaws related to natural resource management and observed a clear relationship between number of bylaws enacted and the effectiveness of decentralization (Ndegwa and Levy 2004) (see Figure 4.3). Horizontal and vertical linkages and cooperation foster strong and well-functioning institutions (Berkes 2002) (Figure 4.3). *Horizontal linkage* entails cooperation among institutions working at comparable levels of organization. For example, at the community level, there are formal and informal institutions that operate at village level, such as local government councils, religious organizations, and custom institutions and projects. As discussed earlier, the effectiveness of these institutions depends on their capacity to enact and enforce a set of regulations that define their organization. Institutions that are well networked horizontally tend to be more effective.

Figure 4.3—Effect of national-level decentralization policy on enactment of land and water management regulations

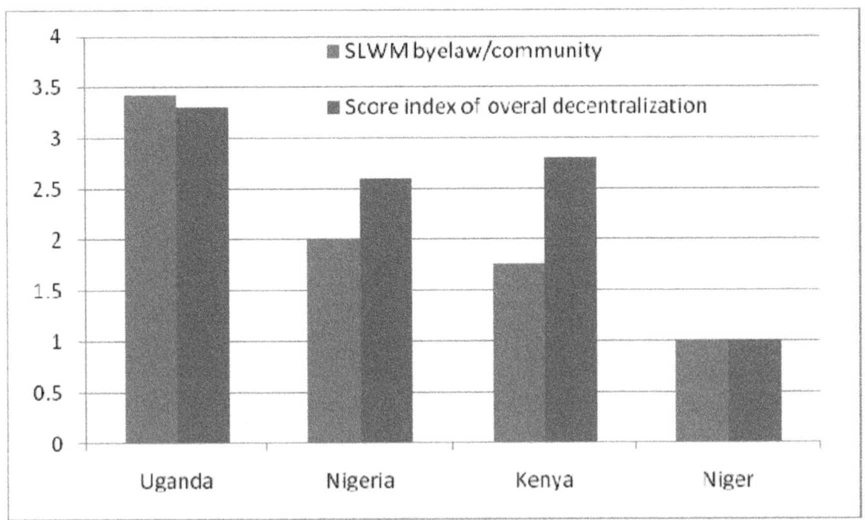

Sources: Overall decentralization from Ndegwa and Levy 2004; SLWM bylaws from World Bank 2010.

Notes: SLWM: sustainable land and water management. Overall decentralization includes 12 performance and structural indicators of decentralization. The larger the index, the greater the performance of decentralization.

For local community institutions to work even more effectively, they need to have strong *vertical linkages* in order to provide the required support for capacity building, legal mandates, and financial resources. Past literature on institutions has emphasized the great advantage of nurturing local institutions to have the right to organize (see, for example, Ostrom 1990). This implies that national level institutions would establish an act that gives mandates and power to local and customary institutions to enact and enforce their own bylaws, which are fully recognized by the upper administrative organization. If such a mandate is missing, development of local institutions will be elusive. Linkage with NGOs supporting community-level institutions is also crucial to build the capacity to organize and take collective action. Figure 4.4 depicts how vertical and horizontal linkages connect institutions at different governmental levels. Key to the vertical linkage is the need for the higher institutions to foster a bottom-up approach that will increase the capacity of local institutions to manage lands and to operate on a long-term basis.

Figure 4.4—Institutional structure with horizontal and vertical linkages

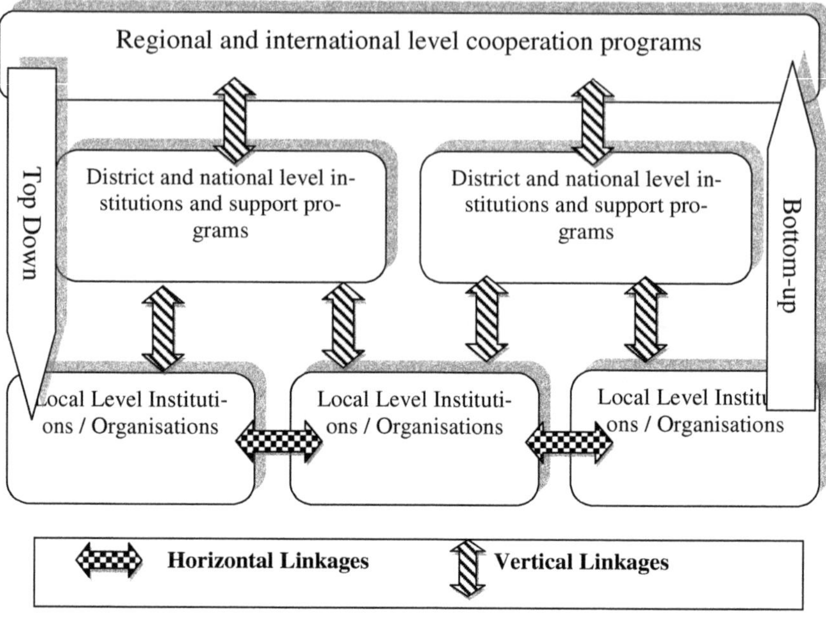

Source: Author's creation.

The bottom-up, community-based natural resource management approach has largely become a panacea to the old poor performance of the centrally, top-down managed natural resource management programs (Berkes 2007). However, an improved top-down approach is required to ensure that ideas from the upper institutions are presented and accepted by lower-level institutions. For example, Qamar (2005) suggested that resource-poor farmers may not demand labor-intensive land management practices with long-term payoffs, such as soil conservation terraces or tree planting. In such cases, there is a need to provide supply-driven extension services, or the national-level government may need to work harder to convince communities to invest in tree planting or to invest in land improvement that they would otherwise not. Likewise, implementation of the Paris Declaration on Aid Effectiveness may also require some prodding and persuasion from developed countries, since poor countries may not see the need for investing in development programs that have long-term payoffs. In such an environment, negotiations have to be made to convince countries or communities to invest in land improvement rather than in investments with immediate payoffs (see Box 4.6).

Box 4.5—Reducing emissions from deforestation and forest degradation and clean development mechanism as examples of vertical integration to change land management patterns

Recent globalization and global change have accentuated the importance of vertical integration (Berkes 2002). Carbon offset and other forms of the payment for ecosystem services (PES) program have created opportunities for building vertical linkages for land management. Horizontal and vertical institutional organization has been identified as one of the major approaches that will allow small land operators to participate in the carbon market and other PES programs (see whole issue of Mountain Forum Bulletin, 2010; Capoor and Ambrosi 2008).

One mechanism to reduce carbon dioxide emissions from deforestation (a form of land degradation) is the Reducing Emissions from Deforestation and Forest Degradation (REDD) scheme, launched by the United Nations Framework Convention on Climate Change (UNFCCC) at the COP-13 (Conference of Parties) meeting in Bali. By using market and financial incentives to reduce the emission of greenhouse gases from forest degradation, REDD offers an opportunity to utilize funding from developed countries to conserve forests in developing countries. Although its original objective was to reduce greenhouse gases, it can also deliver "co-benefits," such as biodiversity conservation and poverty alleviation. Impacts of REDD on the ground still have to be measured, as in most countries (such as Nepal, Nicaragua, and Indonesia), REDD is in a pilot phase.

The Clean Development Mechanism (CDM) is another tool for vertical linkages across countries aiming to reduce carbon dioxide emissions and forest degradation. Countries that have to reduce emissions according to the Kyoto Protocol can buy certified emission reductions (CERs) from countries that have access rights for emissions. Through this exchange, emissions are reduced in areas where reduction is cheaper, while local users gain from forest conservation. However, certification for these programs has high entry barriers, leaving out a number of potential users.

4.4 Summary

The purpose of this section was to embed land users' decision to "degrade" the land or to invest in land conservation into a broader institutional context. *Institutions,* which are defined as the formal and informal rules governing economic production and exchange (North 1991), play a mediating role and can help explain why the same drivers of degradation can lead to different outcomes (for example, population pressure leading to both degradation and improvement). The concept of levels of institutions indicates at which level different schools of economic thought are focusing in their analysis. This presentation of layers, together with the listing of key groups of actors, set the scene for looking more closely at some key institutional arrangements and policies affecting land use decisions.

The most important types of actors involved in land management are direct land users, landowners, governments, custom institutions, industries, beneficiaries of ecosystem services, NGOs, international organizations and development agencies, research institutions, and academia. A global assessment must consider all types of actors and how they interact with each other.

Among the strongest incentives for land users are property right structures, which do not necessarily need to be formalized but which should give sufficient long-term perceptions of security and incentives to invest in land productivity. The discussion on links between poverty reduction and land degradation briefly outlined this important link, with causalities running in both directions. There is potential in reducing poverty through conservation methods; however, adaptation does not always seem to work (Nkonya et al. 2009). Market opportunities also factor strongly in explaining why some actors invest in land conservation (for example, off-farm employment with higher wages; Woelcke 2003), as well as explaining what users choose as an optimal rate of degradation.

Institutions play a key role in shaping the actions or inaction of land users. However, the effectiveness of local institutions is heavily influenced by national-level policies. Decentralization policies, in particular, have played a pivotal role in mandating local institutions to manage natural resources more effectively. In addition, local institutions require capacity strengthening to make them more effective. Establishing horizontal and vertical linkages should improve institutional learning and sustainability and should be considered when designing institutional reform. As policies try to cope with complex multi-equilibrium environmental and social systems, they should aim for flexibility to deal with changing circumstances (for example, climatic variability).

The complexity of the institutional setup presented in this section reflects the complexity of land use decisions and their consequences. Understanding the causes of inaction (or of inappropriate actions) is the key to delivering effective land degradation policies. It also has a direct impact on the costs of action against land degradation by ensuring a path of least resistance to adoption of the measures by land users.

5. Actions for Improvements

This section briefly discusses the different methods used for control of land degradation. It also discusses the effectiveness and profitability of those methods, based on a meta-analysis.

5.1 Soil Erosion

Soil erosion is the most well-known type of degradation due to its visible effects, which have prompted a large number of studies (Nachtergaele et al. 2010). Soil erosion is largely induced by water and wind.

Water-Induced Soil Erosion

El-Swaify et al. (1982) and Junge et al. (2008) identified three major methods of controlling water-induced soil erosion on crop fields:

- Mechanical methods: These include soil and water conservation (SWC) structures, which prevent water movement, and drainage structures, which control passage of runoff. Planting trees, grass strips, and other vegetation also prevents water movement and could be used to enhance SWC structures. For example, Nkonya et al. (2008b) showed that SWC structures reinforced with leguminous plants have lower maintenance costs and are more profitable than when only SWC structures are used.
- Agronomic methods: These include mulching, crop planting pattern (for example, along contour bands), cropping systems (for example, intercropping with crops that have better land cover), planting cover crops, and timing planting to ensure maximum coverage when soils are most vulnerable to water-induced erosion.
- Soil management practices: These include zero tillage, minimum tillage, tie tillage, tillage along contour lines, and so on.

The appropriateness of each practice is dictated by the nature of the water-induced soil erosion, the biophysical characteristics (for example, topography, rainfall quantity, and pattern), and a score of other socioeconomic characteristics—all of which determine their adoption. A combination of these practices is more effective than a single method. It is important to note that these methods will also address wind-induced soil erosion and other forms of land degradation discussed below.

Wind-Induced Soil Erosion

Wind erosion is a major problem in dry areas with poor vegetation. There are no reliable data on wind erosion's impact, due to limited global database (Nachtergaele et al. 2010). Case studies have been done in several countries and regions to assess the impact of wind erosion and the practices to control it. Wind erosion is controlled by establishing windbreaks and by making the soil surface more resistant to wind erosion (Tibke 1988). As with other land management practices (such as integrated soil fertility management, discussed below), wind erosion control is more effective when a combination of control practices is used (Tibke 1988). For control of wind erosion on field crops, Tibke (1988) identified the following five main practices:

1. reducing field width
2. maintaining vegetation residues on the soil surface
3. utilizing stable soil aggregates or clods
4. roughing the land surface
5. leveling the land

A study in the Sahelian region of Africa showed that mulching with crop residues was the most common wind erosion control measure (Sterk 2003). Standing crop residues was 5–10 times more effective in controlling wind erosion than flat crop residues (van Donk 2003). However, due to insufficient quantities of crop residues and competition with livestock, regeneration and exploitation of natural, scattered vegetation was deemed the most promising control strategy in the Sahelian region (Sterk 2003).

Soil Nutrient Depletion

Soil nutrient depletion result from poor land management practice, which, in turn, leads to more outflow of nutrients than inflow. Areas with naturally poor soil fertility, coupled with poor land management, tend to suffer from severe soil nutrient depletion. For example, Natchergaele et al. (2010) showed severe nutrient depletion in sub-Saharan Africa, where both soil fertility and land management practices are generally poor. It is estimated that less than 3 percent of total cropland in Sub-Saharan Africa is under sustainable land and water management practices (Pender 2009).

Recent studies have shown that integrated soil fertility management (ISFM), defined as the judicious manipulation of nutrient stocks and flows from inorganic and organic sources for sustainable agriculture production that fits the socioeconomic environment of farmers (Smaling et al. 1996), is more sustainable than fertilizer or organic soil fertility management practices alone (Vanlauwe et al 2011). Examples of ISFM include practices which combine fertilizer with organic inputs to restore soil nitrogen and organic matter. ISFM promotes judicious use of rock phosphate or inorganic fertilizer to replenish phosphorus and other limiting nutrients. The ISFM approach has become increasingly popular due to its win–win attribute of increasing both crop yield and carbon stock. ISFM also reduces chemical fertilizer application rates and therefore has the potential to reduce environmental pollution that arises from excessive application of fertilizers, which is now common in southern Asia and South America (Phipps and Park 2002; Vanlauwe et al 2011). Some studies have also shown that ISFM is more profitable than the use of fertilizer or organic matter alone. Twomlow, Rusike, and Snapp (2001) found that marginal rates of return (MRR) from a *mucuna*–maize rotation in Malawi were higher than from the use of inorganic fertilizer. Sauer and Tchale (2006) observed similar results in Malawi. Mekuria and Waddington (2002) also found much higher returns from ISFM than from fertilizer or manure alone.

However, other studies have shown ISFM to be less profitable than fertilizer or organic soil fertility management practices alone. For example, in the Machakos district of Kenya, de Jager, Onduru, and Walaga (2004) found a cost–benefit ratio of less than 1 in all trials involving organic and inorganic soil fertility combinations, except one (combining inorganic fertilizer and manure in irrigated maize production)—in the exception, the ratio was only 1.19.

Table 5.1 summarizes the type of land degradation and the solution to address each type. As far as possible, the table also gives some examples of the impacts and profitability of the practices.

Table 5.1—Type of land degradation and their solutions

Type of land degradation/ processes	Solutions	Examples of potential impacts and profitability
Water-induced soil erosion	• Mechanical methods: Soil and water conservation structures; drainage structures • Agronomic methods: Mulching; crop management (cover crops, intercropping, and so on); planting pattern/time • Soil management methods: Minimum tillage or no till; ridge tillage, tie tillage	
Wind-induced soil erosion	• Windbreak and dune stabilization using trees and other vegetative methods • Cover crops in humid or semihumid zones • No till • Rotational grazing and other practices that improve land cover or prevent rotational grazing	• Standing crop residues are 5–10 times more effective in controlling wind erosion than is flattened crop residue (von Donk 2004).
Salinity	• Prevention of salinity • Amelioration using intermittent or continuous leaching • Breeding for saline-resistant crop varieties • Using halophyte crops, trees, and pasture	

Table 5.1—Continued

Type of land degradation/ processes	Solutions	Examples of potential impacts and profitability
Compaction/sealing and crusting	Soil management methods: Periodic deep tillage, controlled farm equipment or livestock traffic, conservation tillage, Agronomic methods: Intercropping or rotational cropping, alternating shallow-root and deep-root crops (for example, maize and beans)	No tillage can save 30–40 percent of labor (Hobbs et al. 2007); gross margin of minimum tillage was 50 percent more than plowing using a hand hoe in Zambia (Haggblade and Tembo 2003).
Loss of biodiversity	Prevention of land use conversions that lead to loss of biodiversity Afforestation and reforestation programs Promotion of diversified cropping and livestock systems	
Soil fertility mining	Integrated soil fertility management (ISFM)	ISFM is more profitable than use of fertilizer or organic soil fertility alone (Doraiswamy et al. 2007; Sauer and Tchale 2006
Soil pollution	Reduced use of agrochemicals Integrated pest management (IPM) Proper use of agrochemicals	IPM is more profitable than conventional plant protection methods (Dasgupta, Meisner, and Wheeler 2007).
Overgrazing	Rotational grazing Planting of more productive fodder Reduction of herd size	Compared to continuous grazing, rotational grazing can increase live weight up to 30 percent in the Sahelian region (World Bank 2011) and up to 65 percent in Canada (Walton, Martinez, and Bailey 1981).
Water shortage	Development of irrigation infrastructure Drought-resistant crop varieties Mulching, and other carbon-sequestering management practices	

Source: Compiled from sources cited.

5.2 Salinity

Salinity is a major problem in semiarid and arid zones (Bot, Nachtergaele, and Young 2000), as well as in irrigated areas with poor drainage. It is estimated that at least 20 percent of all irrigated lands are salt affected (Pitman and Läuchli 2004). Salinity costs global agriculture an estimated $12 billion per year, and this figure is increasing (Pitman and Läuchli 2004). The salinity hot spots in the world include Central Asia and Australia.

Solutions for addressing salinity include the following:

1. *Prevention of salinity buildup by improving the drainage of irrigation systems:* Apart from wasting large volumes of water and contributing to water-borne diseases, poorly drained irrigated areas contribute to water logging and salinity buildup (Bhutta and Smedema 2007; Smedema et al. 2000). Poor leveling could also contribute to salt buildup, due to formation of ponds (FAO 2001). Thus, improving drainage systems will reduce salinity buildup and minimize wastage of water use. Use of fertilizer could also increase salinity. Thus, reduced use of fertilizer or avoidance of the use of some forms of fertilizer (such as sulfate of ammonia) will also reduce the risk of salinity.
2. *Breeding salinity-tolerant crop varieties:* Transgenic varieties of some crops have shown to be saline tolerant (Pitman and Läuchli 2004). Experiments done in Iran have shown that the yields of salt-tolerant wheat, barley, and sorghum varieties could be 50 percent higher than the yields of conventional varieties (Ranjbar et al. 2008).
3. *Use of halophytic plants for economic purposes:* Related to (ii), the use of halophytic, or salt-loving, plants (such as date, palm, barley, and cotton) could help reduce salinity (Natchergaele et al. 2010). Areas with saline soils could be used to grow halophytic crops and fodder (Toderich et al. 2009).
4. *Remediation of saline soils by leaching soluble salts below the root zone:* This practice (Qadir et al. 2006) requires a large amount of water and may not be amenable to areas with limited irrigation water—a common phenomenon in dry areas, where salinity is common. The practice also requires soils with good porosity and a deeper water table to allow efficient leaching. In areas with poor porosity or a shallow water table, soaking the soils and later draining saline water out of the farming area has been done. Mechanical removal of salt crusts has also been used; however, this practice is limited, because it can only remove the crusts on the surface.

Past studies have compared two leaching methods: continuous ponding and intermittent ponding (Qadir et al. 2006). Intermittent ponding (leaching) reduces the water requirement by about one-third the amount required for continuous ponding to remove about 70 percent of soluble salts (Hoffman 1986); however, leaching requires more labor than continuous ponding (Oster 1972). In addition, intermittent leaching combined with mulching reduced evapotranspiration and further improved salt (Carter et al. 1964).

5.3 Compaction

Compaction is a major problem in areas with high livestock population density and in areas where heavy machinery is used for cultivation. Compaction due to livestock pressure is a severe problem in the Sahelian region, the horn of Africa, Central Asia, northeastern Australia, Pakistan, and Afghanistan (Nachtergaele et al. 2010). Compaction due to the use of heavy machinery is severe in the United States, Europe, South America, India, and China (Nachtergaele et al. 2010).

An experiment in Pakistan showed that soil compaction reduced up to 38 percent of wheat yield, largely due to its impact on water and nutrient use efficiencies (Ishaq et al. 2001). Practices used to address compaction include periodic deep tillage and the recently popular conservation agriculture, which uses minimum or zero tillage, control of soil erosion, and water and moisture conservation through the use of crop residues such as mulch, cover crops, and crop rotation (Hobbs 2007). The adoption rate of conservation agriculture is estimated to be about 95 million hectares, which is about 6 percent of the global crop area of 1,527 million hectares (FAOSTAT 2008). However, adoption of conservation agriculture remains limited in developing countries. The United States, Brazil, and Argentina account for 71 percent of the total land area under conservation agriculture (Derpsch 2005).

Reduction of livestock density is one of the methods used to address this problem; however, it has not been successful due to the top-down approach used to implement it (Mwangi and Ostrom 2009; Nori et al. 2008). Therefore, appropriate measures, such as periodic deep plowing, controlled traffic, conservation tillage, and the incorporation of crops with deep tap root systems into the rotation cycle, are necessary to minimize the risks of subsoil compaction.

5.4 Loss of Biodiversity

A recent study on The Economics of Ecosystems and Biodiversity (TEEB) showed that the share of mean species abundance (MSA) in 2000 was below 60 percent of its potential in much of India, northeastern and midwestern United States, northeastern Brazil, India, China, Europe, the Sahelian zone, and Central Asia (TEEB 2010). TEEB predicted that by 2050, the total global loss of MSA would be 15 percent of its 2000 level and that other losses would largely results from managed forests, agricultural areas, natural areas, and grazing areas (Figure 5.1).

Figure 5.1—Global loss of mean species abundance, 2000–2050

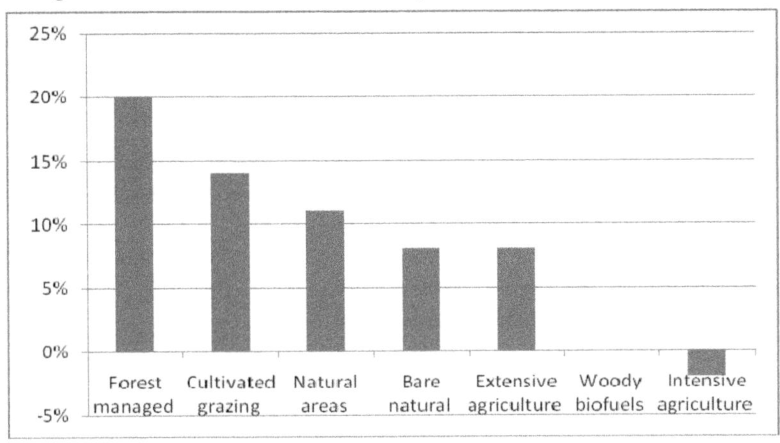

Source: TEEB 2010.

Figure 5.2—Drivers of the loss of mean species abundance at the global level, 2000–2050

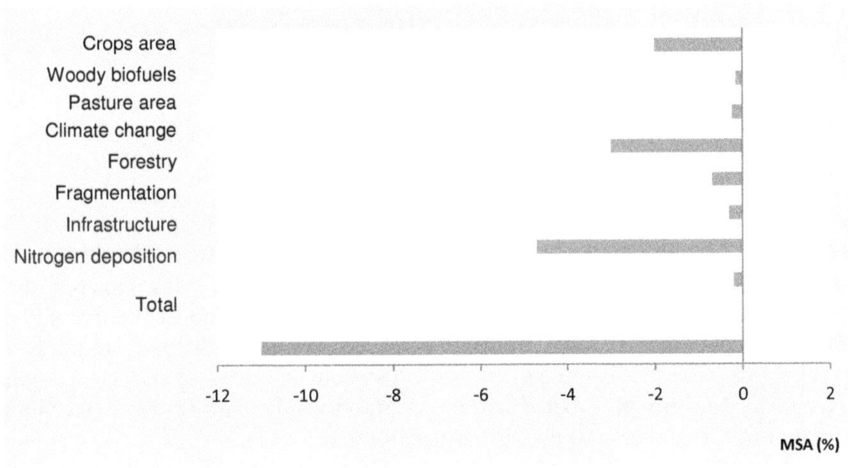

Source: TEEB 2010.
Note: MSA = mean species abundance.

These losses would mainly result from climate change, infrastructure development, pollution, expansion of agricultural areas, and fragmentation (Figure 5.2). Past losses of biodiversity show that conversion of natural habitat into agriculture and other land use types is of major concern for terrestrial biodiversity loss (MA 2005b). Agricultural expansion, which is happening in 70 percent of countries (FAO 2003)—is the largest cause of natural habitat conversion. However, the rate of deforestation is decreasing (Figure 5.3) due to tree planting and protection programs in many countries (MA 2005). Such efforts have helped reduce the rate of biodiversity loss.

Figure 5.3—Forest area as a percentage of total area across regions

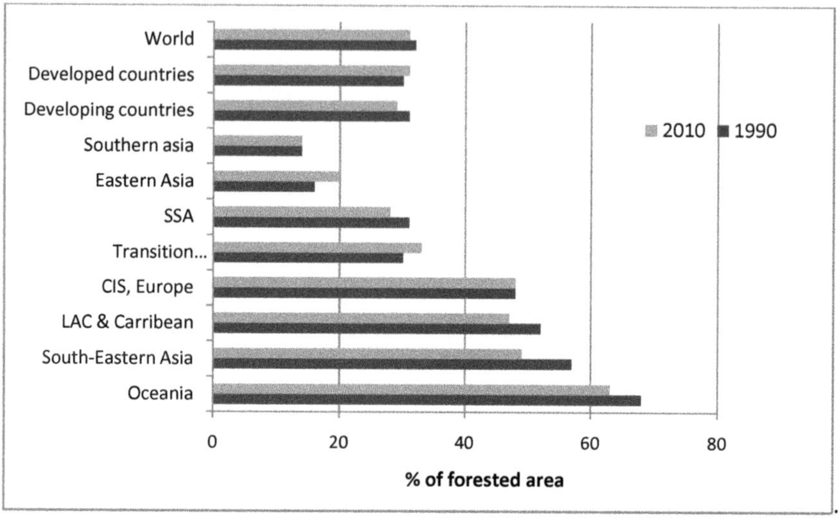

Source: MA 2010.

The TEEB study observed that the conservation costs of biodiversity is cheaper in developing countries than in developed countries. The study also observed that conserving different types of biodiversity was cheaper than protecting just one form of biodiversity. However, the value of biodiversity in developing countries is low, suggesting that protecting biodiversity requires an arrangement to establish payment for ecosystem services. This is especially important, because the beneficiaries of biodiversity are always the local communities, national, and the global community.

5.5 Overgrazing

About one-quarter of Earth's land surface is rangeland that is used by transhumant pastoral communities, which is estimated to consists of about 200 million households and to support about a billion head of cattle, camels, and small ruminants (FAO 2001; Nori et al. 2008. Overgrazing is an especially large problem in the rangelands in arid and semiarid areas. As shown in Table 5.2, livestock population has been declining in almost all rangeland zones due to the expansion of agriculture into rangelands, enclosures that limit the more sustainable transhumant livelihood, and road infrastructure. In Central Asia and Siberia in Russia, however, rangelands have expanded due to decollectivization.

Table 5.2—Status, trend, and drivers of pastoral livestock population

Region	Major livestock type	Status and trend of livestock population
Sub-Saharan Africa	Cattle, camel, sheep, goats	Declining due to advancing agriculture
Mediterranean	Small ruminants	Declining due to enclosure and advancing agriculture
Near East Central Asia	Small ruminants	Declining in some areas due to enclosure and advancing agriculture
India	Camel, cattle, sheep, goats	Declining due to advancing agriculture, but peri-urban livestock production expanding
Central Asia	Yak, camel, horse, sheep, goat	Expanding following decollectivization
Circumpolar	Reindeer	Expanding following decollectivization in Siberia, but under pressure in Scandinavia
North America	Sheep, cattle	Declining with increased enclosure of land and alternative economic opportunities
Andes	Llama, alpaca	Contracting llama production due to expansion of road systems and European-model livestock production, but expansion of alpaca wool production

Source: Blench 1999.

Pastoralists using rangelands have practiced transhumant livelihoods for centuries (see Figure 5.4). Past efforts to address overgrazing have been through campaigns to prevent transhumant livelihoods (Toulmin 2009), which has actually led to even more serious overgrazing due to the large herds owned by pastoralists and the low productivity potential of rangelands.

Figure 5.4—Distribution of pastoralists

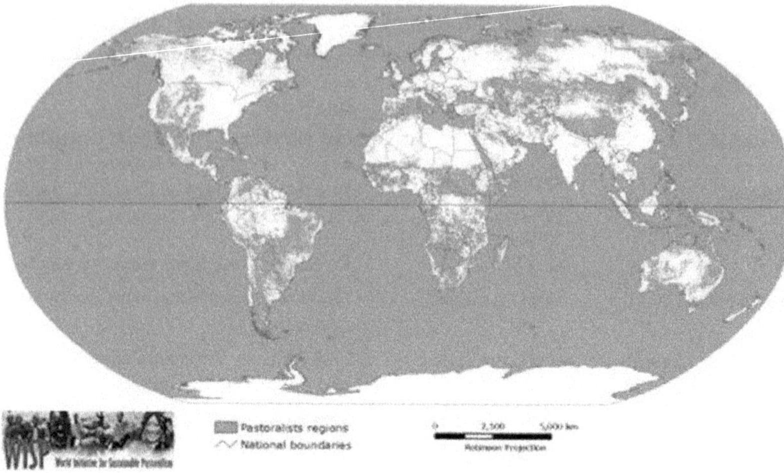

Source: Nori and Davies 2007.

Because the demand for livestock products is increasing, there is a great need to develop rangelands. Involving rangelands in the carbon market is one strategy that could be used to improve rangelands. Another key strategy is to develop local institutions to help manage rangelands. Stronger local institutions will help prevent wildfires, which are common in rangelands, as well as other forms of degradation.

5.6 Addressing water shortage

Due to climate change and other changes, severity of water shortage will increase in arid and semi-arid areas and as discussed above, this will lead to land degraation The following are five important points in the action and mitigation against drought-related water shortage:

Investment in Irrigation and Water Storage

Investment in irrigation infrastructure helps reduce drought-related risks. For example, a study in India showed that investment in dams helped districts downstream of the dams to increase agricultural productivity, reduce vulnerability to rainfall shocks, and reduce poverty (Duflo and Pande 2007). Countries with a heavy reliance on rainfed agriculture are the most vulnerable. Additionally, drought has severe impacts on livelihoods and economic growth. For example, in Sub-Saharan Africa, land area with irrigation potential is 42.5 million hectares, and yet only 12.2 million hectares, or 30 percent of the irrigable area, is irrigated (Foster and Briceño-Garmendia 2010). Globally, irrigated production accounts for 40 percent of total production but occupies only 17 percent of cultivated area (Foster and Briceño-Garmendia 2010). Despite the high frequency of drought, its devastating impact, and the large water availability, Sub-Saharan Africa water storage development is the lowest in the world (Figure 5.5).

Even in areas with well-developed irrigation infrastructure, investment is required to reduce irrigation water loss and improve water use efficiency. This is especially important given the predicted reduction of water due to climate change and the increasing demand for water.

Moisture Conservation and Water Harvesting

These two methods are especially important in semiarid and arid areas. Moisture conservation practices include mulching, incorporating crop residues, planting cover crops, using minimum or zero tillage, and other practices that increase soil carbon. A study in a semiarid site in Kenya showed that mulching increased the length of the growing period from 110 to 113 days (Cooper et al. 2009).

Figure 5.5—Per capita water storage in selected countries and regions

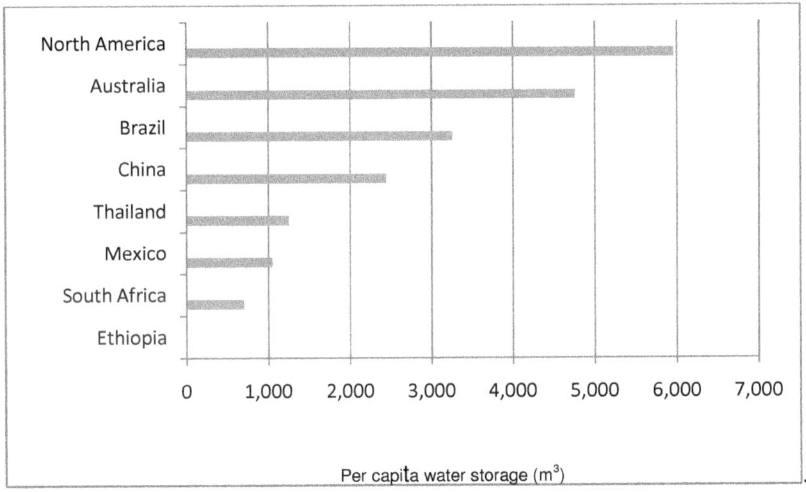

Source: Grey and Sadoff 2006.

In general, mulching and other organic soil-fertility management practices could simultaneously increase crop yield and reduce crop production risks. For example, a study in Uganda showed that soil carbon increased crop yield and reduced yield variance, as shown in Figure 5.6 (Nkonya et al 2011).

Figure 5.6—Relationship between soil carbon and crop yield and yield variance, Uganda

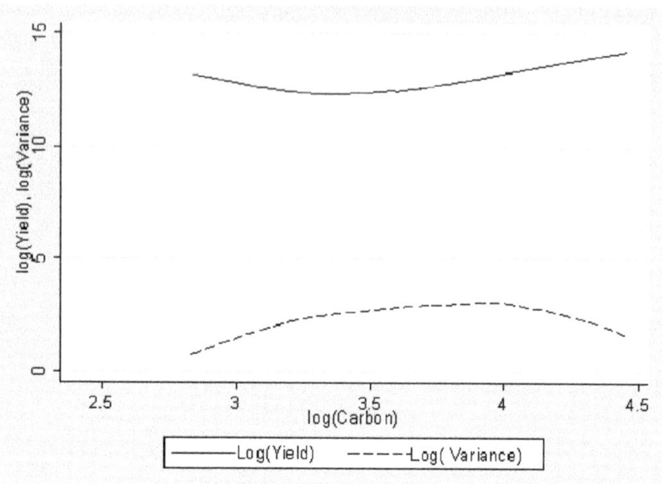

Source: Nkonya et al 2011.

Water-harvesting management practices include tie ridges, water basins for trapping rainwater (for example, the *zai* in west Africa, which was invented by a farmer in semiarid Burkina Faso, is a half-moon planting pit that is filled with crop residues or other forms of plant residues and that traps rainwater and runoff).

6.7 Summary

This section gave an overview of the measures that are suitable to addressing the various types of land degradation. Water-induced soil erosion can be controlled on crop fields by mechanical methods (such as SWC structures), agronomic methods (such as mulching, planning patterns), and soil management techniques (such as zero tillage, minimum tillage). Establishing windbreaks and using planting techniques that make soil surfaces less vulnerable are common ways to reduce the impacts of wind-induced soil erosion. Soil nutrient depletion results from poor land management and leads to more outflow of nutrients than inflow. Soil salinity is a major problem in irrigated areas, with hot spots in Central Asia and Australia. A major management tool against salinity is the appropriate management of water tables. Measures to avoid or mitigate soil compaction and its impacts include improved drainage systems, salt-tolerant crop varieties, and remediation through leaching.

A study by Pender (2009) showed that less than 3 percent of total cropland in Sub-Saharan Africa is under sustainable land and water management practices. Recent studies have shown that integrated soil fertility management (judicious manipulation of nutrient stocks and flows from inorganic and organic sources) is more beneficial to the soil than is the application of fertilizer or organic matter.

Soil compaction may reduce yields quite tremendously—for example, by 38 percent in Pakistan (Ishaq et al. 2001). Whereas compaction due to livestock pressure is a problem in the Sahel, the horn of Africa, Central Asia, northeastern Australia, Pakistan, and Afghanistan, compaction due to heavy machinery occurs predominantly in the United States, Europe, South America, India and China. Practices to address compaction include periodic deep plowing, controlled traffic, conservation tillage, and crops with deep-tap root systems.

The loss of biodiversity was assessed in the TEEB (2010) reports, which predicted a strong reduction in biodiversity (measured as mean species abundance) by 2050. Afforestation, reforestation, protection programs, diversified cropping, and livestock systems are all suitable methods to reduce the species loss.

Overgrazing is a problem in the arid and semiarid rangelands. Transhumant livelihood is a sustainable way of using rangelands that has been threatened by expanding agricultural areas; thus, campaigns to protect transhumant pastoralists have been launched. This efforts need to be accompanied by strong local institutions that help manage rangelands sustainably. Involving rangelands in the carbon market is another strategy to improve rangelands.

To mitigate the impacts of water scarcity, investments in irrigation and water storage, as well as measures for moisture conservation (for example, mulching, incorporation of crop residues) and water harvesting (for example, tie ridges, water basins), will help reduce drought-related risks to people.

6. Case Studies

6.1 Introduction

Five countries were selected to provide an in-depth analysis of the costs of action and inaction. These five countries represent the five major regions in developing countries. The case studies are used to demonstrate the methods discussed in the previous sections. We also discuss some successful case studies in the selected areas that demonstrate the impacts that actions against deforestation and land degradation have had on livelihoods and ecosystem services. We draw lessons from the success stories to illustrate the effectiveness of some of policies and strategies discussed in this study. With per capita income ranging from more than $5,000 in Peru to as low as $400 in Niger (Figure 6.1), the five countries represent a range of economic development. Economic growth in the five countries also varies. Per capita income in all five countries has been increasing, with Peru showing the most robust growth and Niger the least (Figure 6.1).

Figure 6.1—Trend of per capita income in the case study countries

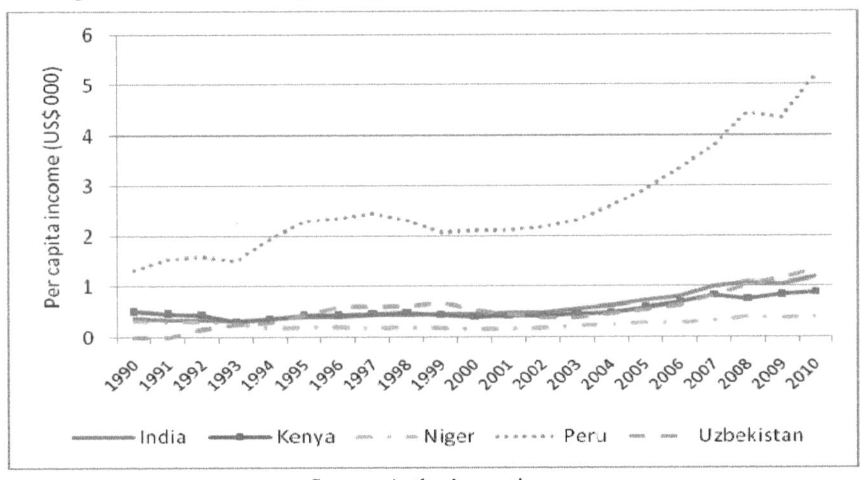

Source: Author's creation.

Land degradation in the five countries is also different. Figure 6.12 shows that Niger has the lowest level of land and biophysical degradation, whereas Uzbekistan has the highest level of both. With the exception of Peru, the case study countries have more than 50 percent of their land area in the arid, semiarid, or hyperarid zone, with per capita arable land area less than 1 hectare

(Table 6.1). This means that the selected countries are largely in the marginal areas.

Figure 6.2—Status of land degradation in the case study countries

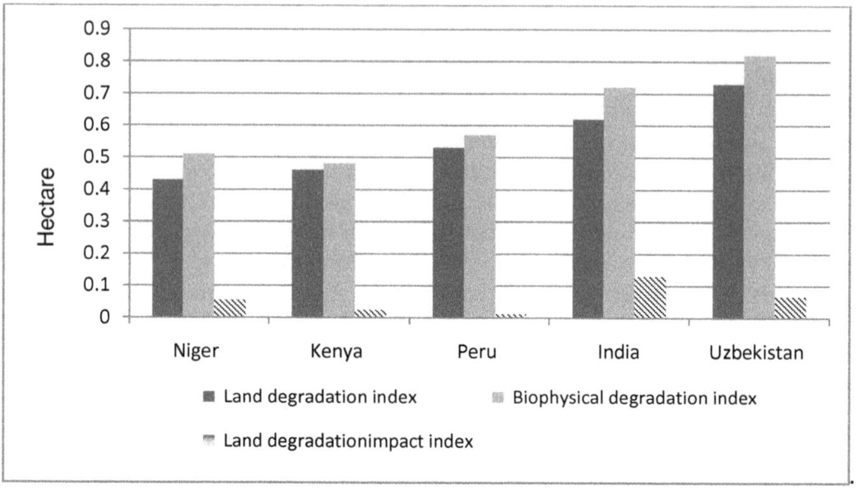

Source: Computed from Natchergaele et al. 2010.

Table 6.1—Land resources and severity of land degradation

Country	Arable land area per capita (ha)	Land area under ASAL (% of total land area)	Soil erosion hazard (% of area)	% of sodicity
India	0.18	72	29	1
Kenya	0.32	80	22	5
Niger	0.44	94	7	1
Peru	< 0.005	34	30	0
Uzbekistan	0.21	91	3	13

Source: TERRASTAT 2010.
Note: ASAL =Arid and semi-arid lands.

6.2 Uzbekistan

Economic Effects of Land Degradation in Uzbekistan

Land degradation is severe in Central Asia, reducing the productivity and threatening the livelihoods of millions of farmers and pastoralists. Major problems include salinity and soil erosion, which affect more than half of the irrigated cropland in some of Central Asian countries. In part due to land degradation, as well as other factors, average yields have declined in many areas by 20–30 percent, contributing to worsening rural poverty and vulnerability. Negative environmental impacts include the drying of the Aral Sea, water and air pollution caused by salinization and erosion, loss of biodiversity, and reduced provision of ecosystem services.

The Republic of Uzbekistan suffers from an environmental degradation crisis that makes sustainable development very challenging. The prevailing arid climate requires that cultivated crops be intensively irrigated. Population growth and the ensuing development of new cropland have caused an increase in water withdrawals, which, in turn, caused a potentially damaging water–salt imbalance (Stulina et al. 2005). Between 40 and 60 percent of irrigated croplands in Central Asia are salt affected or waterlogged (Qadir et al. 2008). The groundwater table is less than 2 meters deep in about one-third of the irrigated lands of Uzbekistan, and in some regions, the share of waterlogged lands is as high as 92 percent (CACILM 2006). Between 1990 and 2001, the area of saline lands in Uzbekistan increased by 33 percent, while the area of highly saline lands more than doubled (Khusamov et al 2009).

Other land degradation problems in irrigated areas include soil erosion, soil compaction, and soil fertility depletion. Especially in sloping and poorly leveled areas, irrigation can be a significant source of water-induced soil erosion. Common cropping practices used in Central Asia, which usually leave exposed soil between rows of cotton or wheat and which involve intensive tillage, expose the soil to significant erosion. Poorly constructed and maintained irrigation and drainage systems, as well as excessive use of irrigation at high rates of flow, also cause significant erosion problems. In Uzbekistan, approximately 800,000 hectares of irrigated cropland are estimated to be subject to serious soil erosion due to poor agricultural practices (poor land leveling, poor irrigation practices, and so on), with annual soil losses of up to 80 tons per hectare of fertile topsoil (CACILM 2006). More than 50 percent of farmland in Uzbekistan is estimated to suffer from serious wind erosion; according to CACILM (2006), soil organic matter has declined by 30–40 percent.

In 2008, IFPRI conducted a study on sustainable land management in Central Asia. In this study, Pender, Mirzabaev and Kato (2009) used a crop modeling software, called the Decision Support System for Agrotechnology Transfer (Jones et al. 2003), to predict wheat and cotton yield responses to alternative levels of nitrogen fertilizer use and, for one of these sites, to reduced tillage practices. Data on fertilization rates, irrigation, agronomic practices, prices, and production costs were used to estimate costs and returns of alternative fertilizer and tillage options. We used the same data set but a different crop modeling tool (CropSyst; Stockle, Donatelli, and Nelson 2003) to assess the economic impact of soil salinity and soil erosion on wheat and cotton[63] production. For both cases, we simulated yields for a 10-year period and used the average for the period to compute the impact on profit.

Salinity

The effects of salinity on crop growth are well documented. Salinity causes a reduction in the amount of water available to the crop due to changes in the soil water's osmotic potential, effects of specific ion toxicity, and increases in the plant's ion concentration (Maas and Hoffman 1977). It is estimated that 10 percent of Uzbekistan is affected by sodicity (TERRASTAT 2010), which is a high concentration of sodium relative to calcium and magnesium (van de Graaff and Patterson 2001).[64] Although the general effects are similar across crops, biological differences cause the same levels of salinity to have different impacts on different crops. The results of our analysis reflect these biological differences. Table 6.2 shows impacts on yields and profit. Figure 6.3 compares the loss in profit for the two crops. It appears that salinity has a greater economic impact on wheat than on cotton.

Table 6.2—Effects of increased salinity on yields and profit for wheat and cotton, Uzbekistan

	Wheat			Cotton		
Simulated levels of salinity (EC_e)	Yield ton/ha	Total revenue/ha	Profit/ha	Yield ton/ha	Total revenue US$/ha	Profit US$/ha
3	4.50	360.00	311.87	2.92	525.60	297.23
7.5	3.71	297.12	248.99	2.90	522.00	293.63
9.5	3.01	241.00	192.86	2.60	467.92	239.55
13	2.23	178.27	130.14	2.11	380.51	152.14

Source: Authors computation from simulation results.
Note: EC_e = electrical conductivity of a saturated soil extract.

63 We looked at the effects of salinity, ceteris paribus. We assumed that farmers continued using the same mix of inputs and made the same managerial decisions.

64 Salinity is a general term referring to a high concentration of types of salts in water or soil; sodicity is a broad term depicting different forms of soil salts. Salinity is one form of sodicity (van de Graaf & Patterson 2001).

Figure 6.3—Profit loss caused by increased salinity, Uzbekistan

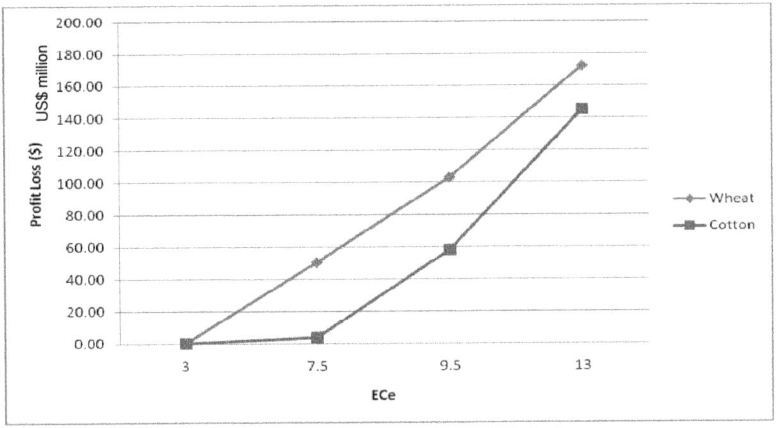

Source: Authors calculations based on simulation results.
Note: EC_e = electrical conductivity of a saturated soil extract

Soil Erosion

We turn now to a simulation of the effects of soil erosion. Erosion adversely affects productivity by reducing infiltration rates, water-holding capacity, nutrients, organic matter, and soil biota. To isolate the effects of erosion, we simulated an increase in slope for the cultivated field and no changes in agronomic practices. We tried to capture the erosion effects by letting the software compute soil loss, reduction in soil depth, and the consequent decreases in yields. Table 6.3 reports the effects of erosion on yields and profit.

Table 6.3—Soil loss, yields, and profit for land in different slope classes, Uzbekistan

	Wheat			Cotton		
Slope class	**Annual loss(ton/ha/yr)**	**Yield (ton/ha)**	**Profit (US$/ha)**	**Annual loss (ton/ha/yr)**	**Yields (ton/ha)**	**Profit (US$/ha)**
No erosion	0	4.50	311.87	0	2.92	297.23
0-2%	1.90	4.48	310.38	0.61	2.92	296.53
2-5%	8.25	4.40	303.93	3.57	2.89	292.44
5-10%	25.49	4.16	284.34	11.03	2.82	279.90
10-15%	58.94	3.62	241.59	25.53	2.67	251.59

Source: Authors computation from simulation results.

Extrapolating the Effects of Land Degradation to the Whole Country

Given the sensitivity to local climate and soil conditions, extrapolating the crop model results from experimental sites to domains beyond the experimental sites is extremely risky. Acknowledging all of the limitations inherent in this attempt, we tried to extend the results to the rest of the country to determine whether some preliminary policy recommendation could be formulated by using this economic approach.

Using IFPRI's crop allocation software Spatial Production Model (SPAM) (You and Wood 2006), we identified the irrigated land on which wheat and cotton are grown and assumed that the entire area is affected by salinity. We computed total profit loss induced by increasing soil salinity from a slightly saline soil ($EC_e = 7.5$) to a moderately saline soil ($EC_e = 9.0$). Similarly, using geographic information system (GIS) software in combination with SPAM, we identified the areas in the different slope classes that had been cultivated with wheat and cotton. We then computed total profit loss caused by erosion. The results are shown in Figure 6.4. Salinity is a major problem, costing the country about $11.21 million annually. Globally, salinity is most severe in Central Asia (Natchergaele et al. 2010a). The economic loss of salinity for wheat and cotton alone is $13.29 million, which is equivalent to the selected crops and is 0.03 percent of the gross domestic product (GDP) of $37.724 billion (IMF 2010).

Figure 6.4—Total profit loss due to land degradation type, Uzbekistan

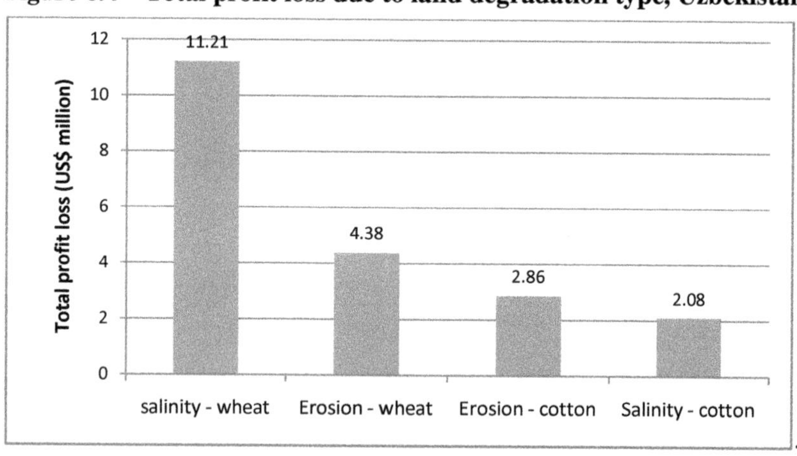

Source: Authors calculations based on simulation results.

The results of our simulation must be taken and interpreted with caution. First of all, as we mentioned, reality on the ground is more complicated than the one modeled. Second, the costs computed do not account for external costs, which, particularly in the case of erosion, can be high. Third, to correctly decide which action should be undertaken, we should also have obtained information on the costs of addressing the different types of degradation. From these simulations, it appears that the most pressing issue is salinity on wheat and that resources should first be devoted to mending this problem.

6.3 Niger

Soil nutrient depletion, overgrazing, salinity in irrigated plots, and deforestation are major problems in Niger. Due to limited rainfall and relatively flat terrain, water-induced soil erosion is limited in Niger. However, wind erosion is a major problem (Sterk 2003). However, we did not estimate the impact of wind erosion due to a lack of data. Niger is among the case study countries with limited use of fertilizer (Figure 6.5). On average, less than 1 kilogram per hectare of nitrogen or phosphorus is used on crop plots.

Figure 6.5—Trend of fertilizer use in Kenya and Niger

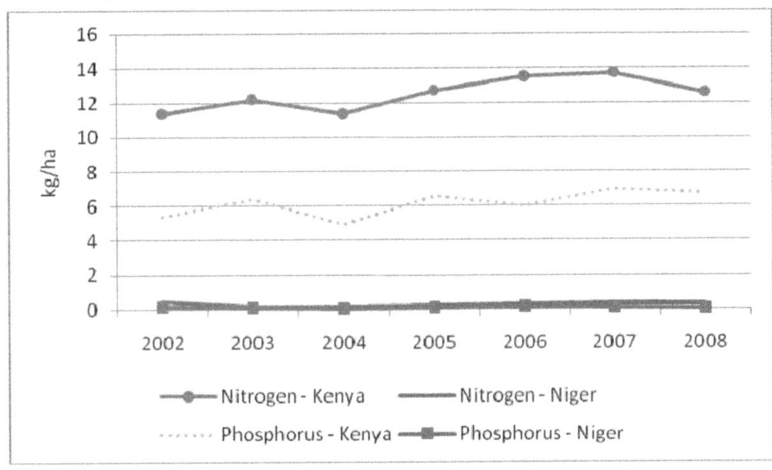

Source: Authors calculations based on simulation results.

As shown in Table 6.4, use of organic soil fertility management practices is also limited. Six percent of households used mulching, and only 1 percent used manure. Adoption rates of all other land management practices were less than 1 percent, which underscores the severity of soil nutrient depletion in Niger. We selected sorghum, millet, and rice production and estimated the loss of profit due to using only crop residues. We compared this land-degrading practice with the use of 40 kilograms of nitrogen per hectare, 1.67 tons of manure per hectare, and the incorporation of 50 percent of crop residues.

Table 6.4—Adoption rates of land management practices in Kenya and Niger

Variable	Kenya	Niger
	% adoption	
Fertilizer and organic soil fertility	33.0	0
Animal manure	68.0	1.0
Fertilizer	36.4	0.1
Improved fallow	4.9	0.6
Crop residue incorporation	34.4	0.1
Mulching	35.2	6.4
Rotational grazing	7.5	0.4
Water harvesting	17.2	0.4

Source: Nkonya et al (2011).

Our results (Figure 6.6), show the loss of profit due to the use of land-degrading practices.

Figure 6.6—Loss of profit due to soil nutrient depletion, Niger

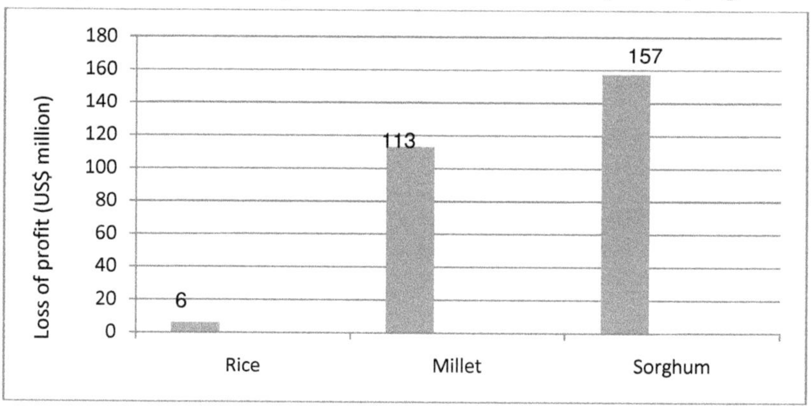

Source: Authors' calculations based on simulation.

The transhumant pastoral system is dominated by the Fulani and the Tuareg (Wane 2005), and overgrazing is a major problem in the Sahelian zone, where the largest share of livestock population is located. Although relatively lower than in other case study countries, the stocking rate in Niger is increasing (Figure 6.7). We estimated the effect of overgrazing for Niger using Erosion Productivity Impact Calculator (EPIC) simulation model. The results show that overgrazing reduces forage yield by 32 percent. A study in the United States found that rotational grazing, as compared with continuous grazing, increased beef gain per unit area by 35–61 percent and the profit from milk by 61 percent

(Henning et al. 2000). Walton, Martinez, and Bailey (1981) also found a 63.5 percent weight gain of cows due to rotational grazing. To obtain a conservative estimate, we assumed that overgrazing reduces carcass live weight and milk production by 32 percent.

Figure 6.7—Trend of livestock units per pasture area in case study countries

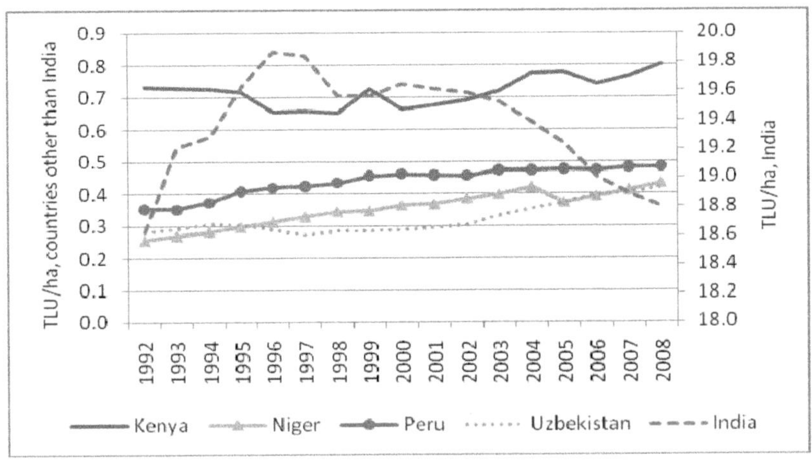

Source: FAOSTAT.

Notes: TLU = tropical livestock unit. TLU is based on a standard animal with a live weight of 250 kilograms. Conversion factor to TLU for livestock: cow = 0.9 TLU; goat or sheep = 0.20 TLU (Defoer et al. 2000).

Costs of Action and Inaction

We evaluated the cost of action and inaction at the farm level. The cost of action is the cost the farmer will incur in addressing land degradation, whereas the cost of inaction is the loss the farmer will incur due to land degradation. In the case of salinity, the cost of action is the cost of water and labor required for leaching. The cost of inaction is the benefit lost due to salinity. This cost is obtained by determining the difference between the net present value (NPV) of practices with desalinization and the NPV without desalinization. Figure 6.8 shows that the cost of action is only about 10 percent of the cost of inaction per hectare, which indicates the high cost that farmers experience by not addressing the salinity problem. We also examined the costs of action and inaction to control overgrazing. Simulation results showed that overgrazing leads to a 22 percent reduction of fodder productivity and a loss of profitability amounting to $1,156 per household with 50 tropical livestock units (TLUs)[65] (Figure 6.9).

Figure 6.10, which depicts the loss of profit as a percentage of GDP, shows that for the selected enterprises alone, Niger loses about 8 percent of its GDP due to land degradation. The results underscore the large cost of inaction to address land degradation.

Figure 6.8—Cost of action and inaction to control salinity in a rice and onion rotation, Niger (US$/ha)

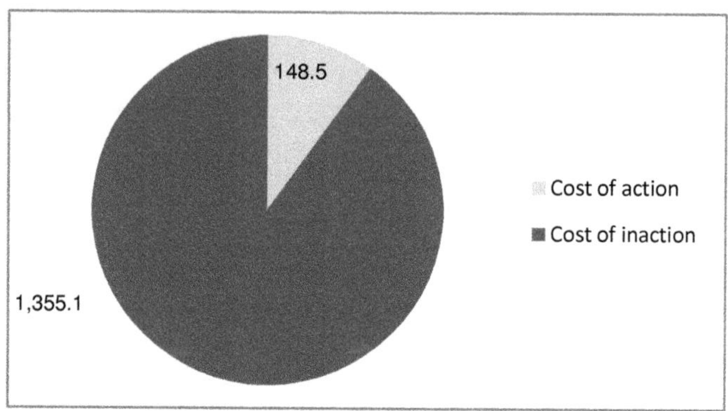

Source: Authors' calculations based on simulation results.
Note: The net present value includes both rice and onion.

65 A standard animal with a live weight of 250 kilograms is called a tropical livestock unit (TLU). The conversion factor to TLU for livestock: cow = 0.9 TLU; goat or sheep = 0.20 TLU (Defoer et al. 2000).

Figure 6.9—Cost of action and inaction to control overgrazing, Niger (US$/household with 50 TLU)

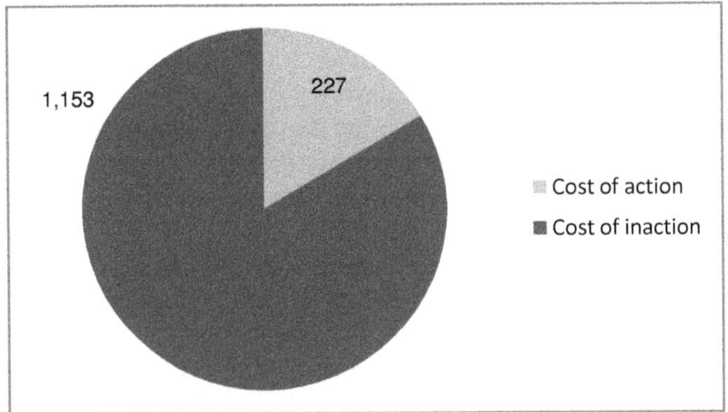

Source: Authors' calculations based on simulation results.
Note: TLU = tropical livestock unit.

Figure 6.10—Loss of profit as a percentage of GDP, Niger

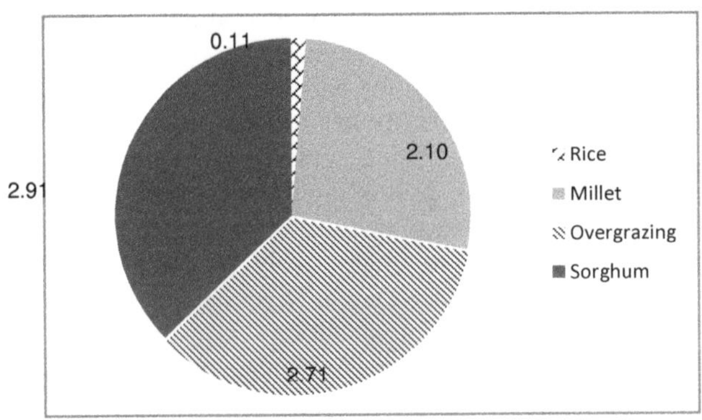

Source: Authors' calculations.

Regreening of the Sahel in Niger

Niger is one of the Sub-Saharan Africa countries that has achieved remarkable land rehabilitation programs. The government and development partners have invested in land management programs, because the majority of the population heavily depends on the land. The Special Program of the President, the Projet de Gestion des Ressources Naturelles (Natural Resources Management Program), and more than 50 other programs have been promoted by the government, nongovernmental organizations (NGOs), and donors since the early 1980s (World Bank 2009). In addition to these investments, the government also revised its institutions and passed the rural code in 1993, which gave customary leaders more powers to manage land and encouraged them to plant and protect trees and to benefit from such efforts without government intervention. The forest policy gave landholders the tenure rights to trees that they planted or protected (Yatich et al. 2008; World Bank 2009). In addition, the government promoted contract farming in state-controlled forests (Yatich et al. 2008). These changes contributed to the sense of ownership and economic incentives that the communities needed in order to participate in protecting the forests. Sales of forest products also helped farmers cope with the country's risky agricultural production.

These policy changes and investments have led to significant recovery of the Sahelian regions where they were implemented. For example, villages where the Projet Intégré Keita; Projet de Développement Rural de Maradi was operating were found to be much greener than what could be explained by just a change in rainfall (Herrmann, Anyamba, and Tucker 2005; Adam et al. 2006; Reij, Tappan, and Smale 2008). In total, tree planting and protection have led to the rehabilitation of 3 million hectares (Adam et al. 2006).

Other important factors also explain this remarkable success. The drought of the 1970s–1980s, which led to a loss of vegetation, created a new value for trees. Following the drought, collection of firewood and water took a day-long task for women. People responded to this challenge by protecting growing trees instead of cutting them, as had been the case in the past. The tree scarcity also affected the livestock sector, especially in northern Niger, where trees are used as fodder during the dry season. Hence, tree scarcity significantly affected the livelihoods of rural communities, prompting them to change from land clearing to tree protection. The Niger government also responded to this challenge. In the 1970s, the government started to aggressively promote tree protection and planting. For example, Independence Day in Niger became National Tree Day.

The NGOs and religious organizations also helped significantly in building the capacity of local institutions to manage natural resources. They also helped mobilize communities to plant and protect trees. For example, the farmer-managed natural regeneration (FMNR)—in which communities protect or plant new trees and harvest fuelwood, fodder, nitrogen fixation from leguminous trees, windbreaks, and other ecosystem benefits—was initiated by a religious organization (Reij, Tappan, and Smale 2008). The authors estimated that villages with FMNR had 10–20 times more trees than they had had before FMNR started. Consistent with Bai et al. (2008a), higher tree density was found in villages with high population density (Reij, Tappan, and Smale 2008).

The lessons we can draw from Niger are the institutional vertical (rural code) and horizontal (grassroots NGOs and religious organizations) linkages, which gave local communities the mandates and the capacity to manage natural resources. The rural code provision, which allowed communities to benefit from their tree planting or protection efforts, also created strong incentives to farmers to invest their limited resources. All these complementary conditions fostered the successful tree programs in Niger.

6.4 Peru

Soil nutrient depletion in Peru is moderate, because the use of nitrogen and phosphorus is considerably large and has shown an upward trend (Figure 6.11).[66] Hence we will not evaluate the soil nutrient depletion problem. Soil erosion remains a major problem in Peru's Andean region, which covers about 30 percent of the country, whereas salinity is a problem in the irrigated crops of the arid and semiarid coastal region, which covers 34 percent of the country. Posthumus and de Graaf (2005) showed that soil erosion reduces maize yield by 2 percent on plots with slope of 1–5 percent. The cost of establishing terraces was estimated to be \$364 per hectare, whereas NPV of plots with terraces—computed after netting out the NPV if a plot did not have terraces—was \$984 per hectare (Figure 6.12). The cost of action is actually lower if we consider that establishing bench terraces is a long-term investment, which further shows that the cost of action is much lower than the cost of inaction.

Figure 6.11—Trend of nitrogen and phosphorus use, Peru

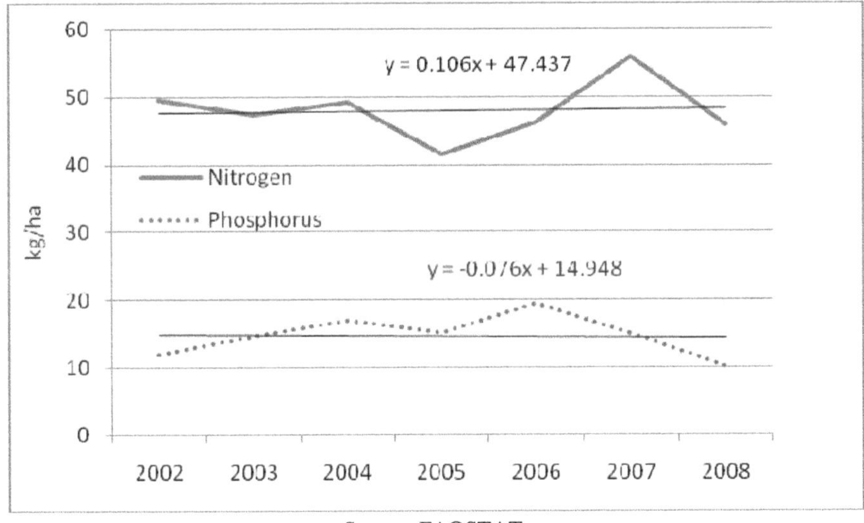

Source: FAOSTAT.

66 Except for 2008, when fertilizer prices abruptly increased, leading to a decline in both nitrogen and phosphorus use.

Figure 6.12—Cost of action and inaction (US$/ha) of soil erosion on maize plots in the Andean region, Peru

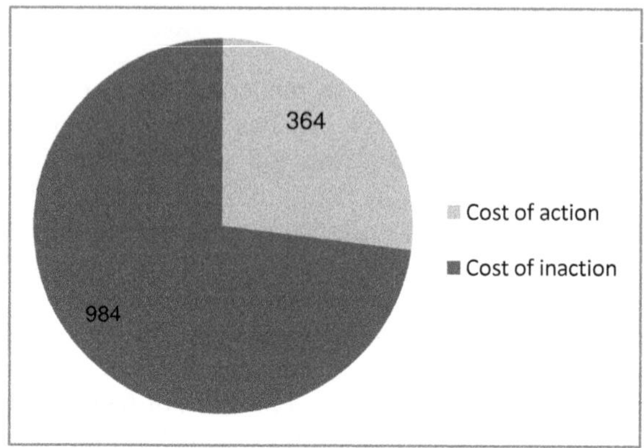

Source: Posthumus and de Graaf (2005).

Notes: Cost of action is the establishment cost of terraces, and cost of inaction is the loss of profit (NPV calculated after netting the NPV that the farmer will get if he or she did establish terraces).

The cost of salinity in Peru was also evaluated, using rice as a case study. Rice yield in Peru is among the highest in the world. For example, whereas the average paddy rice yield in India was 3.2 tons per hectare for 2005–2009, it was more than twice that (7.15 tons per hectare) in Peru (FAOSTAT 2009). However, salinity has a large impact on rice yield. Crop simulation results showed that salinity reduced rice yield by 22 percent in Peru, which leads to a loss of $402 (Figure 6.13). As discussed earlier, salinity could be controlled by staggered leaching, an action that involves more use of water and labor for leaching. The cost of desalinization in Peru was $69 (Figure 6.13), which is only 17 percent of the cost of not taking action to control salinity.

Figure 6.13—Cost of action and inaction to address salinity, India and Peru

	Cost of action	Cost of inaction
India	108	176
Peru	69	402

Cost (US$/ha)

Source: Authors' calculations from simulation results.

Notes: Cost of action includes water for leaching ($100 per hectare in India and $50 per hectare in Peru) and three-day labor costs for leaching ($6.25 per day in Peru and $2.25 per day in India).

Success Stories of Land Management in Peru

Until only recently, Peru's natural resource management had been highly centralized (Anderson and Ostrom 2008), which meant that local communities did not have an opportunity to develop the capacity to locally manage natural resources. For example, the Peruvian government does not give mandates for municipalities to formulate bylaws for natural resource management (NRM), nor does it permit municipalities to raise taxes or transfer funds for NRM (Anderson and Ostrom 2008). The authors concluded that decentralization alone does not guarantee better NRM; rather local institutions have a strong influence on NRM (Anderson and Ostrom 2008). Specifically, interaction between municipalities and local institutions is the key to better NRM (Anderson and Ostrom 2008).

One of the examples of the self-initiation of communities to manage natural resources is community tourism. A good example is the Posada Amazonas, which is a joint tourism agreement between an independent tour company (Rainforest Expeditions) and an indigenous community (Infierno) resident in the rainforest. The community is responsible for protecting the forests and providing tour guide services. The independent tour company provides marketing and transportation services to the tourists. Since the project started, the income of the participating community has increased by 70 percent (Häusler 2009). What is different for this project is its collaboration with a private tour company instead of external funding from the government or donors (Kiss 2004).

Similarly, a new clean development mechanism (CDM) started in Piura in 2009. This CDM plans to use the community-based forest management approach to reforest 8,980 hectares of land in the degraded dry areas of Piura. The project will increase the biodiversity in the arid and semiarid areas of Piura, where overgrazing and deforestation have depleted biodiversity. The reforestation program will still be adapted to silvopasture[67] systems and family orchards in the Piura area and will provide fodder for livestock. In addition, the CDM project is supporting family orchards in order to increase the community's interest in planting and protecting trees and to strengthen the perception that the forests belong to the communities. AIDER (Asociación para la Investigación y el Desarrollo Integral) is also providing support to enhance the capacity of communities to manage natural resources.

Discussion with communities during this study showed that even before the CDM project started, communities in northwestern Piura took deliberate actions to promote agroforestry and silvopasture, which was especially important in the dry forests in the Lambayeque region. AIDER and the community developed a silvopasture strategy in which rotational grazing and other pasture management practices were promoted. AIDER also played a key role in raising the communities' environmental awareness and helped build collective natural resource management. In addition, AIDER promoted the establishment of vaccination and other animal health services. Even though the communities still do not have a mandate to enact bylaws due to the weak decentralization in Peru, the communities' increased awareness helped them to build a more sustainable silvopasture system, which has led to recovery of the arid and semiarid areas of Piura. AIDER tested the sustainability of their program by leaving the communities to operate without support; many of the villages have continued to operate the silvopasture systems more sustainably than before. With the help of AIDER, the communities were able to plan for the conservation of dry forests, from which they benefited from nontimber forest products (NTFPs), such as fruits from the *algarrobo* tree. The boiled fruits of the algarrobo are rich in minerals and sugars and are used to make a*lgarrobina*, a staple food in Peru's arid and semiarid areas. The fruits also are ingredients for livestock feed. The AIDER study showed that the value derived from algarrobo NTFP was greater than the value of charcoal made from the same tree.

These case studies in Peru further demonstrate the importance of involving communities in managing and benefiting from natural resources. It also shows the role played by NGOs in enhancing the capacity of local communities to manage natural resources.

---

67 Silvopasture is a production system which combines forage and forest production.

6.5 India

The nature of land degradation in India is different from what we see in the Sub-Saharan African countries. India is among the countries that benefited from the Green Revolution. Agricultural productivity in India has generally been increasing, due to the increasing use of fertilizer and improved crop varieties. As Table 6.5 shows, the total factor productivity (TFP) of more than 50 percent of the major crops increased from 1970 to 2000, due to the increasing use of fertilizer and other inputs. Nitrogen and phosphorus use has been increasing, respectively, by 4 kilograms per hectare and 2 kilograms per hectare each year from 2002 to 2008 (Figure 6.14). Such high fertilizer use makes soil nutrient depletion a smaller problem.

Table 6.5—Trend of total factor productivity growth of major crops, India

		Declining TFP	Annual TFP growth <1%	Annual TFP growth >1%
Paddy (rice)	1971-86	30.5	25.9	43.6
	1987-00	15.0	32.8	52.2
Wheat	1971-86	10.3	17.3	72.4
	1987-00	2.8	74.7	22.5
Coarse cereals	1971-86	19.8	9.6	70.5
	1987-00	60.2	9.8	30.1
Pulses	1971-86	42.8	36.6	20.5
	1987-00	69.2	26.6	4.2
Oilseeds	1971-86	35.6	18.3	46.1
	1987-00	28.3	10.6	61.1
Sugarcane	1971-86	20.3	61.0	18.6
	1987-00	90.9	5.4	3.7
Fibers	1971-86	53.8	7.2	39.0
	1987-00	32.5	1.4	66.1
Vegetables	1971-86 1987-00	27.5	27.5	72.5

Source: Kumar and Mittal 2006.
Note: TFP = total factor productivity

Figure 6.14—Trend of nitrogen and fertilizer use in India, 2002–2008

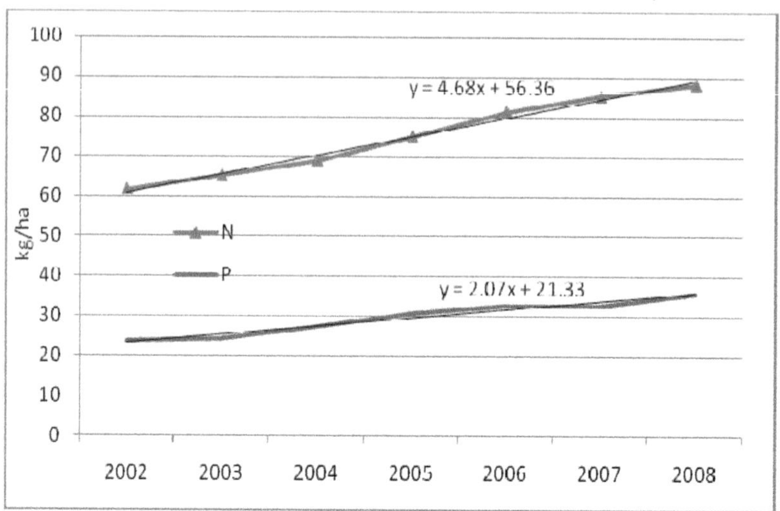

Source: Calculated from FAOSTAT data.

Overexploitation of groundwater is one of India's major environmental problems. Irrigation accounts for approximately 63 percent of total cereal production in India, and groundwater accounted for 45 percent of the 567 cubic kilometers of irrigation water used in 2000 (de Fraiture, Giordano, and Liao 2008; Kumar, Singh, and Sharma 2005). Wheat and rice are the major irrigated cereals; other cereal crops are largely rainfed (Kumar, Singh and Sharma 2005). Salinity is also becoming an increasingly big problem for irrigated crops. It is estimated that about 2 percent of cropped area in India has salinity problem (TERRASTAT 2010). Based on crop simulation models used in this study, salinity reduces crop rice yield by about 22 percent. The cost of action includes the additional cost of desalinization, which, as discussed previously, involves staggered leaching of salts. The cost of irrigation water in India varies from $0 to as high as $470 per hectare in Gujarat (Cornish et al. 2004). We estimated the cost of action (desalinization) to be about $127 per hectare (Figure 6.13). As seen in Niger, the cost of action is smaller than the cost of inaction, suggesting the profit incentive is not the reason for inaction.

Community Watershed Management and its Impact on the Water Table in Tamil Nadu

Rising water due to poor drainage has been one of the challenges of agricultural water in India (Boumans et al. 1988). A study done in Tamil Nadu evaluated the impact of community-based watershed management through *Panchayati Raj* institutions (customary governance institutions), local user groups, and NGOs. Results show that community-based watershed management in Tamil Nadu lowered the water table, increased perenniality of water wells, and increased the availability of water for livestock and domestic use (Kuppannan and Devarajulu 2009). This finding is consistent with other studies that have shown successful community-based natural resource management in India and elsewhere (see, for example, Kerr 2007; Ostrom and Nagendra 2006). The findings are also consistent with the discussion in the institutional section, in which we argued for the importance of local institutions in managing natural resources. The example of India illustrates the importance of participatory and bottom-up approaches, which places natural resource management into the hands of local institutions and communities. A review by Darghouth et al. (2008) shows that participatory watershed management was successful when the programs were of common interest to the community, were flexible, and were a mechanism for capacity building and empowerment of local communities.

As a result of the success of community-based watershed management in India, the government has adopted policies that give mandates to communities to manage watershed issues (Darghouth et al. 2008). However, community-based watershed management has not been effective in managing larger areas of watersheds (Darghouth et al. 2008) or where culturally or economically diverse communities are involved (Kerr 2007). This finding suggests the need for creating well-coordinated vertical and horizontal linkages that will address complex watershed management scenarios, thus further illustrating the argument discussed in Section 4.

Agroforestry Practices and Renewable Energy Programs

India is one of a few countries that has seen a significant improvement in rainfed agriculture. Bai et al. (2008b) showed improvement in rainfed cropland and pastures in western India. Such an improvement is evidence of the great effort the country has put into improving agricultural productivity. A contributing factor to the increased normalized difference vegetation index (NDVI) in rainfed agriculture is the adoption of agroforestry, which has been a traditional practice in India (Pandey 2007). Agroforestry trees in India are found on about 17 million hectares of land (Pandey 2007), equivalent to about 10 percent of India's agricultural area (FAOSTAT 2008). India is one of the leading producers of *jatropha*, a crop that can grow on highly degraded soils and in arid areas. Jatropha has been used to reclaim 85,000 hectares of degraded land (ICRAF 2008) in northern India. In addition, jatropha production on highly degraded lands has helped lift people out of poverty. With an initial estimate of $650 per person, beneficiaries of a project in northern India earned on average $1,200 from sales of jatropha seeds only three years after the initial investment (ICRAF 2008). Targeting degraded lands is one of the key features of this project and could lead to the reclamation of about 30 million hectares of severely degraded land in India (ICRAF 2008).

Similarly, some cities in India have been providing incentives for the use of solar energy to heat water. India spends about 45 percent of export earnings on energy imports (UNEP 2011); but the country has been working hard to increase production of domestic energy (which includes the jatropha production program discussed above). India is currently one of the leading countries in the production and consumption of renewable energy in the world. Investment in renewable energy increased from $46 billion in 2004 to $173 billion in 2008 (UNEP 2011); non–Organization for Economic Cooperation and Development (OECD) countries—in particular, Brazil, China, and India—accounted for 40 percent of this growth. In 2008, India was the sixth country in the world to produce renewable energy (UN Data 2009). One strategy that India is using to promote the use of renewable energy is property tax rebates for those who use solar water heaters, and a number of cities in India have adopted this strategy. The government's innovative incentive mechanism of providing tax breaks demonstrates that the country could achieve significant milestones in reducing consumption of fuelwood and other sources of energy used for heating and lighting.

Investment in Natural Resource and Guaranteeing Employment for the Poorest

India enacted the National Rural Employment Guarantee Act in 2006. Under this social protection act, participants are given a guarantee of employment for at least 100 days (UNEP 2011). About 84 percent of the public works under this program have been directed to water conservation, irrigation, and land investments. It is estimated that the program has provided three billion workdays and benefited 58 million households (UNEP 2011). Even though there have been challenges with such programs in India and elsewhere (Deshingkar, Johnson, and Farrington 2005), they have shown to be a win–win public investment, creating employment, reducing poverty, and enhancing the land and water resources that are so important for the poor (UNEP 2011).

6.6 Kenya

Overgrazing, soil nutrient depletion, and soil erosion are major problems in Kenya. Figure 6.7 shows that Kenya has the highest number of livestock per unit area. Overgrazing is a problem in pastoral areas, which account for 60 percent of the livestock population in Kenya (Davies 2007). Losses due to overgrazing are estimated to be about 1 percent of the gross domestic product (GDP) (Table 6.6). However, this is a conservative estimate, as it does not include losses of biodiversity, prevention of soil erosion, and so on. For example, using the contingent evaluation approach, Davies (2007) estimated that each household in Kenyan pastoral communities obtain $334 from climate amelioration by rangelands and $103 from rangeland biodiversity.

Table 6.6—Economic loss due to overgrazing, Kenya

Type of loss	Loss
Reduction of livestock products (US$ million)	23.33
Animal wasting (US$ million)	192.60
Total loss	215.92
2009 GDP (US$ billion)[1]	30.14
Loss as percentage of GDP	0.72

Source: IMF World Economic Outlook data. 2010.

Costs of Action and Inaction of Soil Nutrient Depletion

As shown in Table 6.3 and Figure 6.15, the use of soil fertility management practices is limited, even though they are relatively higher than in Niger and many other Sub-Saharan African countries. Results of the impact of soil nutrient mining were estimated using crop simulation models. As was the case for Niger, we compared the cost of action of preventing soil nutrient by using 40 kilograms of nitrogen per hectare, 1.67 tons per hectare of manure, and the incorporation of 50 percent of crop residues. We compared this practice with the incorporation of 100 percent crop residues only. We estimated the cost of inaction as the difference of profit between the two practices. The results (Figure 6.15) show that the cost of action to address soil fertility mining in maize and rice is smaller than the cost of inaction; however, for sorghum the cost of action is greater. The higher cost of action for sorghum underscores the weak response of sorghum to soil fertility inputs like fertilizer. The results suggest that for some crops, organic soil fertility management is more profitable than the use of integrated soil fertility management (ISFM), which uses fertilizer and organic inputs and which is currently being promoted as a sustainable land management practice (Vanlauwe 2007).

Figure 6.15—Costs of action and inaction to address soil nutrient mining for selected crops, Kenya

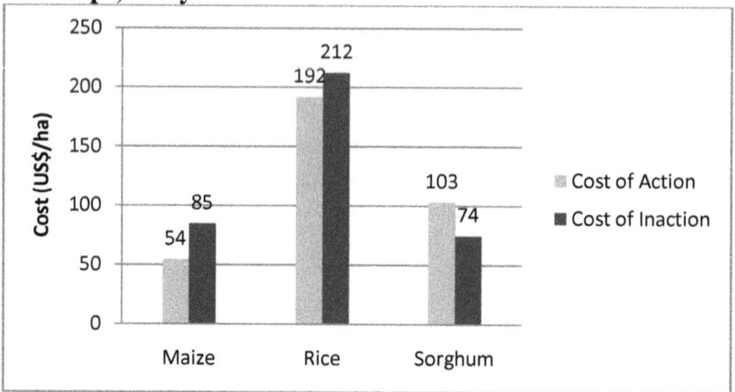

Source: Calculated from simulation results.

Off-Site Impact of Soil Erosion

Nkonya et al. (2008b) estimated the off-site cost of soil erosion in Kenya using a potable water dam supplying water to Nairobi. Box 6.1 shows that the cost of siltation of the water dam was about $127 million, or about $1,000 per square kilometer of the watershed area. The high off-site cost of siltation underscores the need for designing locally based payment for ecosystem services, in which land users upstream could be given an incentive by downstream communities and businesses to prevent soil erosion. In broader terms, cooperation between upstream and downstream communities is likely to enhance better land management practices.

Success Stories of Kenya

Kenya is one of the few African countries in which agricultural policies have been conducive to the agricultural sector. Kenya spent about 1.43 percent of agricultural GDP on research and development, which is almost twice Sub-Saharan Africa's level of 0.7 percent (Beintema and Stads 2004; Flaherty et al. 2008). Crop breeding and research in other management practices contributed to the increased crop yield (Smale and Jayne 2008). Cereal productivity increased about 0.042 tons per hectare, which is exactly twice the growth in comparison to the next highest yield (excluding South Africa) increase in West Africa (Figure 6.16). The open market policies followed by Kenya since its independence have also fostered competitive markets, which have provided incentives for farmers to invest in agriculture.

Box 6.1—Sasumua water treatment plant (Nairobi City Water and Sewerage Company Ltd.)

The Sasumua water treatment plant supplies 1.972 million cubic meters of water per month to Nairobi during the rainy season. The Sasumua dam receives water from the Chania river, which has a catchment of about 128 square kilometers. Deforestation and other forms of land degradation upstream have led to an increase of sedimentation in the Sasumua dam, which has increased the dredging and purification costs. The Sasumua water treatment has seen decreasing water quality and has taken steps to address some of these problems:

1. Higher turbidity due to solids, such as soil, crop residues, animal droppings, and so on—This is addressed by using alum, a coagulant that helps purify water.
2. Higher bacterial count—This is addressed by chlorination.
3. pH increases—The treatment plant does not address this problem.
4. Coloration
5. Agrochemicals loading—This problem is not addressed.

Comparing treatment costs of 1995 and 2005, water treatment for the wet season lasting seven months has changed, as shown in the table below.

Type of cost	**Additional cost (US$)**
Alum (coagulant)	74,499
Chlorination	2,129
Sludge removal (back-wash)	5,525
Dredging costs	44,872
Total additional cost	127,025

Figure 6.16—Cereal yield trend in Kenya compared to Sub-Saharan Africa's regional yields

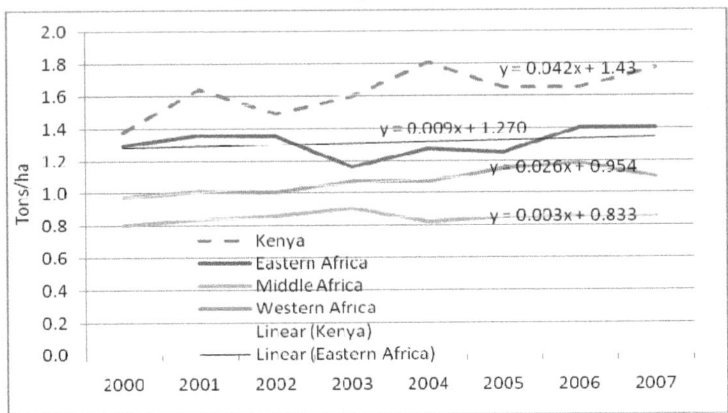

Source: FAOSTAT.

With virtually no fertilizer subsidy, adoption rates of soil fertility management practices in Kenya have been quite high compared with other countries. For example, Kenya applied the sixth largest amount of nitrogen in Sub-Saharan Africa (Figure 6.17). Of interest to us is the strategy the Kenyan government followed to improve access to fertilizer. Instead of focusing on fertilizer subsidies, the government opted for promoting the free market and reducing transaction costs. As a result, the government has been displaced from the fertilizer market by private importers and distributors in most parts of the country (Kherallah et al. 2002). Moreover, Kenya has witnessed rapid investment in private fertilizer distribution networks (Ariga, Jayne, and Nyoro 2006). In 1996, there were 10–12 private importers, 500 distributors and wholesalers, and about 5,000 fertilizer retailers, whereas by 2000, the number of retailers reached between 7,000 and 8,000 (Kherallah et al. 2002). It was only in 2007 that Kenya introduced fertilizer subsidies; however, unlike other Sub-Saharan African countries (for example, Nigeria and Malawi), its subsidy program was administered through the private fertilizer traders. The policies discussed above set Kenya as an example of using private traders to administer fertilizer subsidies and to invest in other programs that reduce transaction costs.

Figure 6.17—Nitrogen fertilizer application per hectare in Sub-Saharan African countries

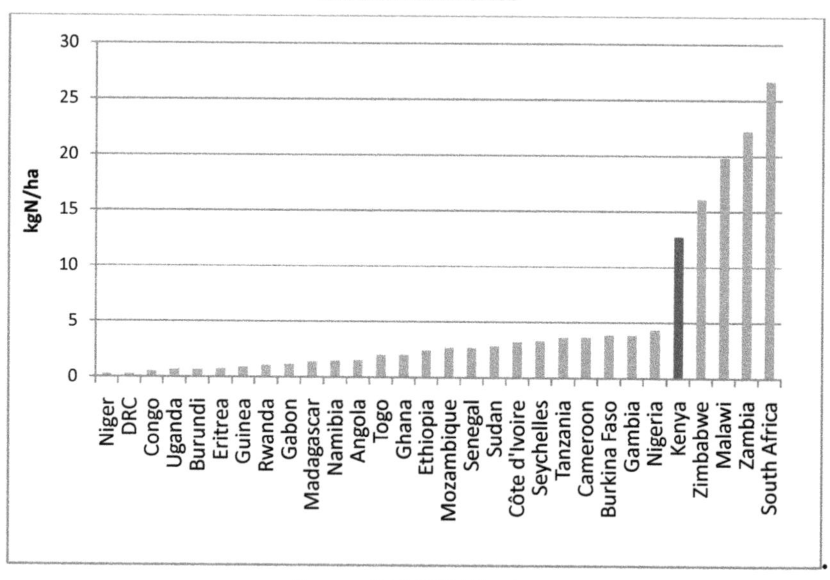

Source: FASOSTAT data.

More People, Less Erosion

A long-term study in Machakos, Kenya, revealed that despite an increase in population density in Machakos, the extent of soil erosion decreased due to the investment in controlling soil erosion (Tiffen, Mortimore, and Gichuki 1994). The major drivers of such a success story are the proximity of the district to the Nairobi market, good infrastructure, and other supportive services that provided incentives for farmers to invest in land improvement (Boyd and Slaymaker 2000). The presence of a large number of NGOs, international research institutions, and international agriculture in Kenya has also contributed to the development and promotion of natural resource management. For example, NGOs and other civil societies have been working with the government to promote soil conservation and fertility measures. These organizations have complemented the public extension program and have brought innovative approaches for promoting sustainable land management practices (World Bank 2010). Advanced large-scale farming in Kenya has also led to significant growth in the use of improved land management practices.

Lessons from the Kenyan study are unique, as they show the impact of national-level policies on land management practices. The main conclusion is that policies that support agriculture and land investments have a significant impact on land management practices at the farm level. Even though Kenya's performance on decentralization is weak (as discussed earlier), its open market policies and strong support of research and development have had a favorable outcome on land management and agricultural productivity. The country will definitely enhance land user and public land investments if it also revises its policies to give greater mandate to local governments to manage their natural resources. The recent constitutional reforms are pointing in this direction.

6.7 Assessment of Forest Ecosystem Goods and Services

We also evaluated the costs and benefits of forest area changes in the case study countries. We used the value of forest ecosystem services per hectare of forest. This assessment does not take into account the benefit from the alternative use of forests and, therefore, cannot be regarded as an evaluation of the costs and benefits per se. We used results of a global study that evaluated forest ecosystem services, which included the value of tropical forests (CBD 2001). Biodiversity and climate regulations accounted for the largest value (Table. 6.7). Climate regulation services include carbon sequestration, wind barriers, and an avoidance of sea rise and crop damage. Biodiversity information value is only for genetic information. We assumed an average value of $400 of biodiversity services, which is a low value range.

Table 6.7—Value of tropical forest ecosystem goods and services

Ecosystem service	Value (US$/ha)
a. Timber – sustainable harvesting	300-2,660
b. Fuelwood	40
c. Nontimber forest products	0-100
d. Recreation	2-470
e. Watershed regulation	15-850
f. Climate benefit	360-2,200
g. Biodiversity (genetic information only)	0-3,000
Total minimum value (a + b + d + f + g)[a]	1,117

Source: CBD 2001

Note: [a] Assuming the value of biodiversity is $400.

To obtain a conservative estimate, the minimum value of each ecosystem service was used. Some of the ecological services listed in the table are not mutually exclusive. For example, watershed benefits are not mutually exclusive with climate benefits. Hence, we only took the sum of ecosystem goods and services that are mutually exclusive to obtain a value of about $1,117 per hectare of forest. Figure 6.18 shows that India's and Uzbekistan's forests increased, which resulted in an increase in the value of ecosystem goods and services equivalent to 0.03 percent of the GDP. Kenya, Niger, and Peru experienced deforestation, and their loss of forest ecosystem goods and services ranged from 0.05 to 0.26 percent of the GDP.

Figure 6.18—Average change in forest area and its value

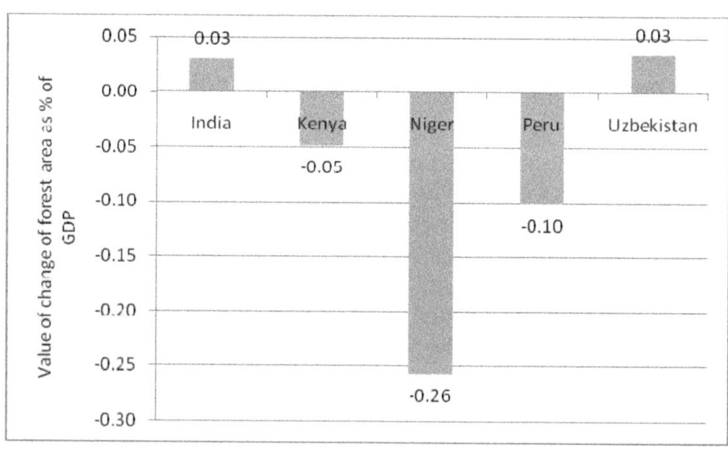

Source: FAOSTAT data.

7. Partnership Concept

The review of past studies on land degradation (LD) clearly shows that there has been a weak link in studies from biophysical scientists and socioeconomists. A need to strengthen this link has increasingly gained attention due to the little attention given to land degradation by policymakers and other decisionmakers. The review also shows that collaboration among biophysical scientists and socioeconomists has been increasing due to the increasing need to determine the causes and economic impacts of LD and the economic benefits of the prevention or reversal of LD (Croitoru and Sarraf 2010; Buenemann et al. 2011).

Based on reviews of the causes and impacts of LD, implementation of programs to address LD also requires a strong collaboration across various actors discussed earlier. As has been seen, causes of LD and its impacts span from the farm level to the global level. In addition, satellite imagery data—which can be collected at a global level at an affordable cost and in a short time—do not capture some important biophysical and socioeconomic data. Thus, there is a need to conduct site-specific measurements using case studies that are selected to represent all major ecosystems and human characteristics. Results of such local studies could then be extrapolated to comparable areas using geographic information system (GIS) and other spatial techniques.

In this section, we propose a partnership that could be used to design a research and communication strategy, implementation of research recommendations, and a monitoring and evaluation strategy.

7.1 An Institutional Setup for the Global Assessment of ELD

As part of a concept for the policy process of setting up a global economics of LD (ELD) initiative, it is important to note the lessons learned from existing global assessments. We suggest as the starting point to refer to the processes and structures behind The Economics of Ecosystems and Biodiversity (TEEB) studies and the operation of the Intergovernmental Panel on Climate Change (IPCC).

The main issue raised in both cases, in terms of translating the scientific work into policy actions, is linked to the credibility of the scientific output. The issue of scientific credibility is
linked to

- the separation between the scientific body and the political body, and
- the quality and diversity (in terms of fields of competence) of the science community involved.

In addressing the former, there must be clear guidelines from the beginning of the scientific work that establish the complete independence of the science body from the policy body in terms of how results are achieved and what these results are. The political body's *droit de regard* ("right to monitor") of the work on the scientific body must be limited to guidance on translation of the evidence-based science into policy-relevant results.

The scientific quality of the work produced is best assessed through a peer review process. For the trust of the political body, the peer review committee must be selected with the political body's approval and must also be formally appointed by it. This would further ensure that the scientific work, once reviewed, will be fully endorsed by the policy body.

Finally, a global assessment of ELD can only be credible if it is truly global in its coverage. Thus, the scientific and policy partners must represent as wide a selection of regions and countries as is possible and manageable. In particular, the scientific work must be undertaken jointly by scientists and research organizations in the developed and developing regions of the world.

We present the conceptual framework of the partnership in Figure 7.1.

Figure 7.1—Institutional setup for a global assessment of ELD in a cost-of-action-versus-inaction framework

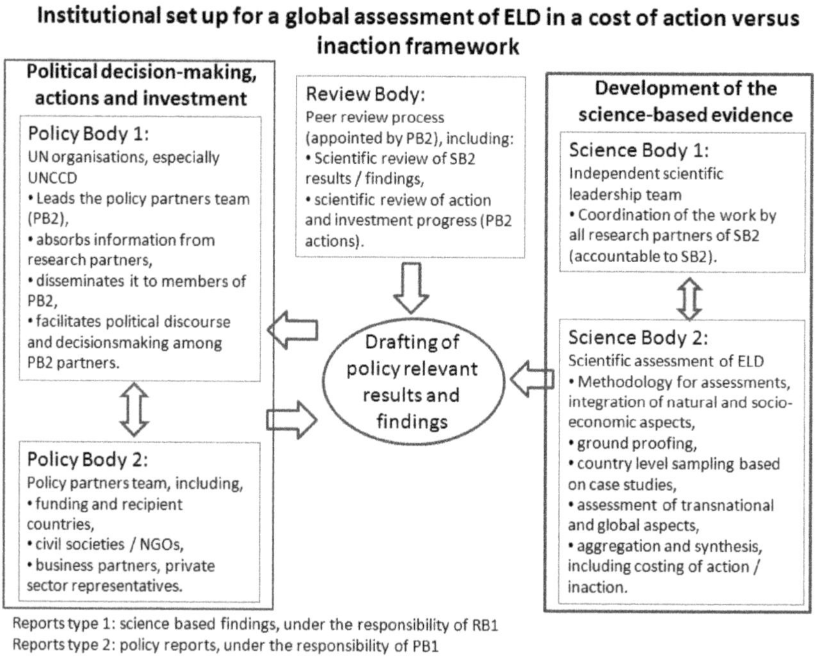

Source: Author's compilation

Notes: UNCCD = United Nations Convention to Combat Desertification; NGO = nongovernmental organization; ELD = Economics of Land Degradation

The Political Partnership

The political partnership is further divided into two groups: the political partnership at the global level (PB1) and the political partnership at the country level (PB2). PB1, which may guide the research process and mobilize resources for conducting ELD, may comprise UN organizations with United Nations Convention to Combat Desertification (UNCCD) playing a key role. The first step of PB1 is to coordinate partners and to define both the problem and the need for taking action to address the problem (this coordination and definition of the problem has already started through the inception of this study).

The PB1 and PB2 will also be responsible for implementing the recommendations made by the ELD study. Implementation of ELD recommendations will be done from the global level to the community and individual land user level. To help properly plan this implementation, the ELD study must provide empirical information showing the extent, severity, and impact of land degradation; who is affected by it; and what is required to prevent reverse land degradation.

The Science Partnership

The science partnership is also at two levels. The first level (SB1) is the core science partnership, which will guide and coordinate the ELD study at global level. This team will come from institutions that have done significant research on the global assessment of LD. The second level (SB2) comprises the communities of scientists from specialized land degradation assessment institutions, as well as scientists working at the regional or country level. The ELD study should show the causes of land degradation and the costs and benefits of preventing or mitigating land degradation. All this should be done spatially in order to gain a full understanding of the distribution of impact, benefits, and responsibility of action. Such analysis will help provide site-specific recommendations to the PB1 and PB2.

The Peer Review System

For the ELD assessment to be credible, it must be peer reviewed. The peer reviewers should be drawn from scientists who are not participating in the study. The review process should also include an evaluation of the implementation of the ELD study's recommendations.

7.2 Processes and Approaches for the Scientific Assessment of ELD

Given that LD is a widespread problem and financial resources are scarce, we propose an approach that allows decisionmakers to prioritize their actions. This approach is based on the idea that LD must be acted against in areas where it shows the highest impacts on human well-being. Such areas can be identified according to several indicators. A list of such indicators needs to be developed within a comprehensive assessment of ELD. Parameters could include the following:

- State of the environment and ecosystem services
- Vulnerability of the local population (that is, dependence on land resources and their benefits)
- Density and size of the local population
- Likely impacts of actions for improvement now and in the future (this links to the state of the local environment and its evolution)
- Likely evolution of LD in the absence of appropriate action

We suggest an economic analysis combined with land degradation monitoring methods, such as remote sensing, GIS, and modeling based on the cost-of-action-versus-inaction framework. The approach should calculate all costs associated with a certain action against degradation and then compare it with the costs of inaction, which consist of a business-as-usual behavior. To include all costs under the two scenarios, it is highly recommended the approach take into account all benefits associated with all ecosystem services, as well as all the direct and indirect costs that variations in the delivery of such ecosystem benefits carry through the environment and the society—notably, via markets and their mechanisms. For example, increased land degradation causes a variation in the provisional services of an agroecosystem, which in turn could lead to a decrease in crop yields, which would have a direct impact on farmers (income or food security), as well as indirect impacts on regional food security through regional food markets and prices, and thus on the costs of food insecurity to the national government. Such mechanisms—especially the analysis of the indirect costs of land degradation—need to be researched further. The linkages among LD, climate variability, and volatility in food markets and their impacts on poverty and human-well being are among the most urgent research priorities. Box 7.1 provides an overview of the proposed approach.

Relying on the evaluation of specific costs at the case-study level and on the more global mapping of the indicators above, this framework allows a global assessment with cost components that are actually rooted in the local conditions. However, there are clear difficulties to applying the methods in the empirical research (case studies) in order to measure the direct costs of LD in a global assessment. More research is required to refine scenario definitions, upscale from site-specific case studies to the national level, and determine how to comprehensively capture the direct and indirect costs of LD.

Because land degradation is a process whose impacts are apparent by relative comparisons over time and because action against it often has delayed results, the assessments need to include a (dynamic) time dimension (for example, through long-term measurements or trend analysis). This is also important to consider in the context of climate change and its complex linkages with LD.

Box 7.1—Summarized stepwise approach

- Assess the status of LD by means of satellite-based imagery on land cover and trends. Add key ecological and social variables on the global level to help identify suitable locations where LD has a high impact on humans and the environment.
- Identify as case studies suitable locations, regions, and countries in terms of different land use systems, land degradation types, and severity, as well as affected ecosystem services and the socioeconomic embedding. Ground-trusting the satellite data is essential to make sure the locations match the expectations. Selection of the case studies could be based on the Food and Agriculture Organization's (FAO's) land use systems classification.
- Perform in-depth analysis of the situation on the ground:
    - Identify and measure important ecosystem services, their level of degradation, and their change over time given continued land use to complement the "economic" approach with a set of well-chosen physical indicators. This step includes on- and off-site assessments of land use on ecosystem services.
    - Perform an economic valuation using appropriate valuation techniques, depending on the kind of services.[68] The concept of total economic value (TEV) provides guidance on all values of ecosystems that need to be assessed.
    - Identify suitable sustainable land management (SLM) options and their impact on the provision of ecosystem services over time, as well as all relevant costs of the installment and maintenance of the SLM.
    - Identify technological innovations and their costs and benefits (for example, plant and animal breeding, extension, water use efficiency).
- Essential for valuation is the identification of the impacts of LD—and, hence, of a loss in ecosystem services—on livelihood, economy, and social developments, as well as to take into consideration the given institutional arrangements in which those impacts take place and to determine the causes and feedback loops between them. This step helps ensure that costs are estimated from society's point of view and include all social impacts and externalities.
- In other words, all direct and indirect effects associated with LD are considered.[69] To that effect:
    - Describe the status quo situation of land use, institutional and policy arrangements and livelihoods to generate the inaction scenario.
    - Identify policy measures and institutions that address the causes of land degradation. Assess all related costs of building and implementing institutions and policies. Combine SLM and institutions and policies into action scenarios.

---

68 Provisioning services, such as agricultural output, have been widely analyzed in literature (for example, using replacement cost or productivity change approaches). Recently, more attention has been paid to other ecosystem functions that cannot be evaluated as straightforwardly as provision services. Some economic valuation techniques exist to assess these functions (for example, contingent valuation, choice experiments, and so on). However, there are still shortcomings associated with the methods used for measuring and valuing complex ecosystem services.

69 The assessment of some of these effects is likely to require sophisticated modeling techniques using bioeconomic models.

Box 7.1—Continued

- Perform a dynamic cost–benefit analysis of inaction and action scenarios to get the present value of the (net) costs of actions and the (net) cost of inaction.
- Upscale the cost estimates for the representative case studies to the global level, based on the results of satellite imagery.[70]
- Determine the cost of action versus inaction for different action scenarios worldwide.
- Determine the cost of immediate action against LD versus the costs of delayed action worldwide.

The assessment of the causes and impacts of LD will require collecting data that cannot be captured by satellite or ground-truthing satellite data. For example, the assessment of the drivers of the adoption of sustainable land management practices requires interviews with land users (Nkonya et al. 2010). However, collection of socioeconomic data is expensive, which, as discussed below, raises the need to partner with bureaus of statistics, which collect household socioeconomic data.

Finally, the global assessment of LD and its costs should systematically review the institutions that determine the actions of stakeholders involved, from local land users to national governments and international institutions. An increasing body of information and knowledge on the causes, effects, and costs of LD must ultimately translate into appropriate action plans.

7.3 Which Type of Capacity Must Be Represented in the Scientific Body?

Past studies have tended to be specialized and have, in the process, produced rigorously analyzed results that have informed researchers in other disciplines. Table 7.1, though not exhaustive, gives some examples of organizations that have performed research, and the strengths that can be tapped in the new partnership. As Table 7.1 shows, a large number of organizations are conducting research on terrestrial ecosystems. The collection of the biophysical data of terrestrial ecosystems is especially significant, even with the weaknesses discussed below. The collection of socioeconomic data is weak, because only a few institutions are collecting such socioeconomic data, mainly because a great deal of important socioeconomic data cannot be captured. Partnership with institutions that collect socioeconomic data will help reduce data collection costs and institutionalize ELD. This partnership will fully use the existing data collection resources. A large number of countries routinely conduct household surveys (see, for example, the list of household surveys conducted worldwide at the International Household Survey Network[71]). These socioeconomic house-

70 Based on satellite imagery, complemented by national statistics and research, values can be transferred to other comparable sites by the benefit transfer method.

71 www.internationalsurveynetwork.org

hold surveys have recently been georeferencing respondents, making it easy to link such data to the biophysical data collected by satellite. However, the household survey data do not include good biophysical data. Thus, one objective of partnering with the bureau of statistics would be to include survey modules that collect good biophysical data.

Table 7.1—Partnerships and the role and strengths of partners in conducting global ELD

	Major organizations that have conducted global assessments	Strengths	Prospective role[1]
Soil erosion and wind erosion; soil nutrient depletion	FAO, UNCCD, ISRIC, UNEP, Global Mechanism, UNFCCC	Global assessment and long-term biophysical data collection; publicly available data	All roles, with varying degrees of focus
Biodiversity	CBD, GEF, Biodiversity, UNEP, TEEB	As above; TEEB also collected and analyzed economic data.	All roles, with varying degree of focus
Loss of vegetation and other forms of land degradation	U.S. NOAA and other satellites	Global assessment, with high-resolution, long-term data; relatively cheap data collection	5, 7
Cost of environmental degradation	World Bank, UNEP; universities (for example, Wageningen University, Texas AandM, University of Bonn, University of Maryland, Université Catholoquie Lovain/Universiteit Leuven)	Rigorous assessment of case studies; development of theoretical framework of assessment	1, 5, 6, 7
Drought monitoring	World Meteorological Organization	Early warning and forecast; global data	5, 6, 7
Soil health surveillance	AFSIS: 60 sentinel sites; a spatially stratified, hierarchical, randomized sampling site of 100 km^2, representing major ecosystems in Sub-Saharan Africa. Data at http://opendatacommons.org/licenses/odbl/1.0 .	Collection and analysis of land and water management data, which is essential for promoting, protecting, and restoring land, water, and ecosystem health	5, 7
Agriculture and environmental monitoring	Various organizations (see Table 7.2)	Specialized data on anthropogenic ecosystems (agroecosystems) and natural ecosystems. Many indicators used. Many organizations share data freely.	5, 7

Table 7.1—Continued

	Major organizations that have conducted global assessments	Strengths	Prospective role[1]
Desertification studies	French Scientific Committee on Desertification	Provides data, conducts studies, and communicates with French and international policymakers	1, 2, 3, 5, 7
Agricultural monitoring	Observatory for world agricultures (OAM)—Pilot countries covered: Costa Rica, Mali, Niger, Thailand, and Madagascar; FAO GIEWS—Global coverage with a network of 115 governments, 61 NGOs, and many international research institutes, news services, and private-sector organizations	OAM: Equipment, land, fuel, nutrients, water, greenhouse gas emission, production, value chain data, and so on. Collected after 5 years. GIEWS; Food security, drought, disease outbreak; receives economic, political, and agricultural data collected	5, 7
Global terrestrial observing systems (GTOS)	River discharge, water use, groundwater, lakes, snow cover, glaciers, ice caps, ice sheets, permafrost, seasonally frozen ground, albedo, land cover (including vegetation type), fraction of absorbed photosynthetically active radiation (FAPAR), leaf area index (LAI), aboveground biomass, soil carbon, fire disturbance, soil moisture (*See details in table 7.2 on essential climate variable monitoring programs.*)	Part of the GCOS, which is managed by FAO	5, 7
GEO (Group of Earth Observations)	Global monitoring of agricultural production, with emphasis on data that could facilitate higher productivity, reduction of risk, timely and accurate national (subnational) agricultural statistical reporting; forecasting and early-warning systems, global mapping; monitoring and modeling of change in agriculture, land use, socioeconomic, and climate changes.	More than 30 international organizations; currently cochaired by the University of Maryland, the Joint Research Center of the European Commission (Ispra, Italy), and the Institute of Remote Sensing Applications (Beijing, China)	5, 7

Table 7.1—Continued

Socioeconomic characteristics that could explain drivers and impact of land degradation and the prevention of land degradation	Routine national household surveys *(See list at International Household Survey Network at www.internationalsurveynetwork.org.)*; Living Standards Measurement Study—Integrated Surveys on Agriculture (LSMS-ISA), which collects panel data from seven Sub-Saharan African countries (Ethiopia, Malawi, Mali, Niger, Nigeria, Tanzania, and Uganda)	National bureaus of statistics; Bill and Melinda Gates Foundation; support of LSMS-ISA; World Bank	5, 7

Source: Author's compilation.

Notes: 1 = Organization; 2 = Building political will; 3 = Communication and awareness creation; 4 = Fund raising; 5 = Generation and analysis of data; 6 = Implementation of policy recommendations; 7 = Monitoring and evaluation.

ISRIC = World Soil Information; NOAA = National Oceanic and Atmospheric Administration; AFSIS = Africa soil information service (www. AfricaSoils.net); GIEWS = Global Information and Early Warning System; GCOS = Global Climate Observing Systems.

The ELD study will take advantage of the wealth of data and results produced by the organizations and institutions listed in Table 7.2. With such a large number of programs and networks collecting data, coordination and harmonization are required so that the data can be collected in a way that ensures synergies and representativeness and that avoids overlaps and duplications. Ongoing data collection efforts have also offered important lessons that can be used in the ELD—in particular, the experience of and lessons learned from the global terrestrial observing systems (GTOS) apply to other networks and data collection efforts. Latham (2011) observed the following lessons learned by GTOS observations:

- *Sustainability*: Some important data collection and monitoring techniques are implemented by medium- or short-term projects. Given that some important terrestrial ecosystems change slowly, however, medium- and short-term data collection efforts are likely to miss the lagged impacts of land degradation or of land management practices that prevent or mitigate land degradation. However, human and financial resources are required to maintain such long-term data collection and monitoring. Thus, a coordination of efforts (discussed below) could be one of the most important strategies for reducing data collection costs and for enhancing synergies
- *Coordination and interpolability:* To avoid overlaps and duplication and to enhance synergies and representativeness of data collection, there is a need to harmonize and coordinate the large number of organizations collecting data in order to obtain a collection that is representative of major land use types, agroecological zones, and socioeconomic aspects. To enhance interpolation of results, sentinel sites need to be selected such that they are representative of the major ecosystems and socioeconomic characteristics. Selection of data collection sites has been largely influenced by the interest of the organizations that fund the data collection, the people involved in the data collection, and the objectives of the data collection. As a result, many important data collection efforts produce data that cannot be interpolated to the entire world. Sub-Saharan Africa and South East Asia are especially underrepresented in data collection efforts. Such coordination efforts will allow interpolation of the data at a global level, thus enhancing the global ELD study.
- *Integrated approach:* As emphasized earlier, ELD will require global data that cannot be obtained using satellite imagery along. In situ data collection is required to validate satellite data and to collect data that cannot be observed remotely. Modeling is also required to determine the impacts of land degradation and of land management practices used to prevent or mitigate land degradation.

- *Training and capacity building:* The capacity of data collection institutions in developing countries is low and requires training to ensure better data collection. New data collection methods and ever-changing global issues also require regular training, even in middle- and high-income countries, in order to enhance their capacity to collect new types of data and to use new methods and tools.
- *Data accessibility*: There is a need to enhance data availability by increasing access to free data. Currently, there have been increasing efforts to enhance data sharing. For example, several knowledge-generation and knowledge-sharing initiatives have started in recent years, such as FAOSTAT, AQUASTAT, and TERRASTAT; the 1994 UNEP/FAO Digital Chart of the World; the "open GIS;" and others listed in Tables 7.1 and 7.2. It is also important that the data are user friendly.

Table 7.2—Essential climate variable monitoring programs of the terrestrial ecosystem services (part of SB1 and SB2)

Metric	Frequency	Monitoring organizations and institutes	Spatial scale
River discharge	Daily	GTN-H	In situ
Water use	Daily to annual	FAO AQUASTAT	In situ
Groundwater	Monthly	IGRAC plus associates, NASA, ESA, DLR, WMO	In situ
Water level	Daily to monthly	GTN-H plus associates, WMO	Equal
Snow cover	Daily to monthly	GTN-H, NASA, NESDIS, NSIDC, NOAA, WWW/GOS surface synoptic network (depth), national networks	Equal
Glacier cover	Daily to monthly	GTN-G (with WGMS), GLIMS, WGI, NSIDC, IACS, ESA (GlobGlacier)	Annual/ multiannual
Permafrost (below 0 degrees Celsius for two or more years)	Daily to annual	GTN-P plus associates	In situ
Albedo (instantaneous ratio of surface-reflected radiation flux to incident radiation flux over the shortwave spectral domain)	Daily to monthly	WMO, WGCV, BSRN, FLUXNET	Satellite
Land cover	Seasonal to annual	FAO, GTOS/GOFC-GOLD, ESA, NASA, IGBP, GLCN	Satellite

Table 7.2—Continued

Metric	Frequency	Monitoring organizations and institutes	Spatial scale
Fraction of photosynthe-tically active radiation (PAR) absorbed by a vegetation canopy	Daily to annual		In situ
LAI (leaf area index) = one-half the total green leaf area per unit ground surface area	Daily to annual	WGCV, FLUXNET, NASA, GLOBCARBON, JRC-TIP, CYCLOPES, LANDSAF	Satellite
Biomass	Annual to 5 years	FAO Forestry, FLUXNET, ESA, national surveys	Equal
Fire	Daily to annual	ESA, NASA, WGCV, GFIMS, GOFC-GOLD, GFMC	Satellite
Soil carbon	Annual to 5 years	FAO-IIASA world soil map, FLUXNET, national surveys	In situ
Soil moisture	Daily to annual	FLUXNET, WWW/GOS surface synoptic network	In situ

Source: Author's compilation.

Notes: BSRN = Baseline Surface Radiation Network; CYCLOPES = computer algorithms for assessing LAI, fcover (vegetation cover fraction), and FAPAR (see Baret et al. 2009); GFIMS = Global Fire Information Management System; GLIMS = Global Land Ice Measurements from Space; GOFC = Global Observations of Forest Cover; GOLD = Global Observations of Land Cover Dynamics; GLCN = Global Land Cover Network; GTN-G = Global Terrestrial Network for Glaciers; GTN-H = Global Terrestrial Network for Hydrology; GTN-M = Global Terrestrial Network for Mountains; GTN-P = Global Terrestrial Network for Permafrost; GTN-R = Global Terrestrial Network for River Discharge; IACS = International Association of Cryospheric Sciences; IGBP = International Geosphere-Biosphere Programme; IGRAC = International Groundwater Resources Assessment Centre; IIASA = International Institute for Applied Systems Analysis; JRC-TIP = Joint Research Centre Two-Stream Inversion Package; LANDSAF = Land Surface Satellite Analysis Facility; NESDIS = National Environmental Satellite, Data, and Information Service; NSIDC = National Snow and Ice Data Center; WGCV = Working Group on Calibration and Validation; WGI = World Glacier Inventory; WGMS = World Glacier Monitoring Services.

Forming the Research Team of ELD

The research partners for the ELD study should consist of key researchers working on and from the terrestrial ecosystems depicted in the case studies. Table 7.3 gives examples of scientific partners and potential institutions to engage in the scientific process. The scientific team will conduct research in collaboration with the institutions listed in Table 7.1 and Table 7.2.

The research team proposed in Table 7.3 is based on recent work in which the institutions have conducted leading research. Regional balance should also be considered, with partners from the developing world being included, as well as partners whose country is represented in the case studies.

Table 7.3—Example of ELD research partnership team (SB1)

Partner	Activities relevant to ELD and *type of ecosystem service*	Region and specific task on ELD
Ohio State University	Carbon sequestration and land degradation studies—*Forest and land cover*	Global—land degradation and soil carbon
Seoul National University	Strong research programs on land conservation in Asia—*Forest and land cover*	Asia—Biophysical modeling of land degradation
Climatic Research Unit (CRU), University of East Anglia	Leading research on climate change	Global – drought research
University of Bonn (ZEF)	Biophysical and socioeconomic research on global land degradation and land conservation	Global—land degradation and economics of ecosystems
University of Pretoria	Leading role in the 2005 Ecosystem Assessment	Africa—Economics of land degradation
Instituto de Estudios Publicos de la Universidad de Chile	Responsible for the regional Global Environmental Outlook (GEO) report *Agricultural Research in Latin America.* University ranked ninth in Latin America in 2010 (www.webometrics.info/top200_latinamerica.asp).	Latin America—Economics of land degradation
IFPRI	International research	Global—Economics of land degradation
FAO	Leading role in past research on land degradation; currently conducting detailed biophysical and socioeconomic impacts of land degradation and land conservation	Global—Build on past research on land degradation
Hebrew University	Research on desertification	Arid and hyperarid regions—desertification
International Union for Conservation of Nature (IUCN)	Strong emphasis on biodiversity	Global—biodiversity
Comité Scientifique Français de la Désertification (CSFD)	Has conducted studies on the economics of land degradation in selected countries	Africa and Asia—Economics of land degradation
Potsdam Institute for Climatic Impact Research (Germany)	Studies on climatic impacts and land degradation	Global—Land degradation and drought research

Table 7.3—Continued

Partner	**Activities relevant to ELD and *type of ecosystem service***	**Region and specific task on ELD**
International Social Innovation Research Conference (ISRIC)–World Soil Information	Global soil information and land degradation studies	Global—Soil degradation and impacts of conservation practices
M.S. Swaminathan Research Foundation (India)	Sustainable agriculture and rural development research; contributed to Millennium Ecosystem Assessment study	Asia—Land conservation practices, and impacts on sustainable agriculture
UNEP	Led in writing The Economics of Ecosystems and Biodiversity (TEEB) report; has enormous data and resources for environmental management	Global—Cross-cutting issues
Global Mechanism	Research on land degradation	Global—Sustainable land management research
McKinsey and Company	Provides good link to private sector but with strong natural resource research orientation	Global—role of private sector in land management
University of Maryland	Global studies of vegetation and other satellite data analysis	Global—satellite data analysis of vegetation cover
Foundation for Advanced Studies on International Development (FASID; Japan)	Land management and development in Asia and Africa	Asia and Africa—institutions and policies for land management
World Bank	Socioeconomic studies in developing countries; for example, supports household surveys	Global—Socioeconomic studies
United Nations University, Institute for Water, Environment, and Health (UNU-INWEH), and partners, including Stockholm Environment Institute	Land and water management	Global—social economic studies

Table 7.3—Continued

Partner	**Activities relevant to ELD and *type of ecosystem service***	**Region and specific task on ELD**
Centre for Development and Environment, University of Bern	Interdisciplinary research on land degradation and sustainable land management; special program on mountain areas; World Overview of Conservation Approaches and Technologies (WOCAT) member	Global—mountain areas
Indian Council of Agricultural Research (ICAR)	Agricultural ecosystem services	Asia—biophysical and social economic research
Embrapa—The Brazilian Agricultural Research Corporation	Agricultural ecosystem services	Latin America—biophysical and social economic research
Nepal Research Center on Mountain Zones	Agricultural ecosystem services	Asia—biophysical and social economic research
Chinese Academy of Agricultural Sciences (CAAS)	Agricultural ecosystem services	Asia—biophysical and social economic research
Université Cheikh Anta Diop de Dakar (UCAD)	West African Science Service Center on Climate and Adapted Land Use (WASCAL) partner	Sub-Saharan Africa—biophysical and social economic research
Université d'Abomey-Calavi Benin	WASCAL Partner	Sub-Saharan Africa—biophysical and social economic research
University of Ghana	WASCAL Partner	Sub-Saharan Africa—biophysical and social economic research

Source: Author's compilation.

8. Conclusions

Since the publication of the Brundtland Report (*Our Common Future*) in 1987 and the consequent Earth Summit on sustainable development, global attention on natural resource scarcity and degradation has been increasing. This global awareness of natural resource degradation has accelerated because of climate change and rising food and energy prices. In turn, this awareness has led to a growing interest in land investments by the private and public sectors. Despite this interest, however, land degradation has not been comprehensively addressed at the global level or in developing countries. A suitable economic framework that could guide investments and institutional action is lacking. This study aims to overcome this deficiency and to provide a framework for a global assessment based on consideration of the costs of action versus inaction. Thus, a type of Stern Review (Stern 2006) for land degradation (LD)[72] is aimed for on the basis of this study. The urgency of land degradation problems, increased value of land, and new science insights all suggest that the time is ripe for a global assessment of the economics of LD (ELD).

Although climate change has attracted much attention and investment—thanks to the Stern Review (2006), which urged the world to take action now to reverse the adverse impacts of climate change to avoid the costlier delayed response—land investment to prevent or mitigate land degradation has been low. One major reason for this inaction is policymakers' and decisionmakers' limited knowledge of the cost of global land degradation and of its underlying causes. Other than in the case of climate change, this potential slow-onset disaster lacks credible and strong voices.

Because the majority of the poor lives in rural areas—and thus heavily depend on land for their livelihoods—land degradation affects them the most and has high human costs. Furthermore, land degradation affects not only direct land users but also the whole economy. Indirect human and economic costs of land degradation are complex and not well understood yet. Thus, all costs of land degradation must be better understood in order to guide investments in actions to prevent and mitigate it.

This study reviewed the literature on LD with an objective of establishing the state of the art of ELD.

---

72 A list of acronyms is presented at the end of the report, for consultation while reading.

Early Global Assessments of Land Degradation Focused on Dry Areas and a Few Types of Land Degradation but Played a Key Role in Raising Global Awareness.

The global-level assessment of desertification and land degradation started in the 1970s. Early global studies on land degradation largely focused on determining the biophysical forms of land degradation in dry areas. These studies showed an increasing extent and severity of land degradation, albeit focusing on a few forms of land degradation—in particular, soil erosion. Nonetheless, these early studies played a major role in raising global awareness of the severity of land degradation and in helping to formulate global conventions and international and national land management programs. Due to limited technological tools of the time, the early global desertification and land degradation studies relied on expert opinion and were therefore prone to subjective judgment and large errors. For example, the 1977 desertification map reported that 35 percent of the global land area was affected by desertification, and yet the Millennium Ecosystem Assessment (MA 2005a) study showed that only 10–20 percent of the global land area is affected by desertification.

Developments in Remote Sensing and Spatial Technologies Have Opened New Possibilities for Better Assessments of Land Degradation, its Underlying Causes, and its Impacts on Human Welfare.

The development of satellite imagery and other spatial analysis techniques have greatly improved the accuracy and lowered the cost of global assessments of desertification and land degradation, as well as of remotely visible socio-economic characteristics. The first global assessment of land degradation to take advantage of satellite imagery and other spatial technologies was the 1981–2003 Global Assessment of Land Degradation (GLADA). This study differed significantly from past studies in that it assessed land improvement and attempted to analyze the association of changes in both vegetation and the underlying causes of land degradation.

Satellite imagery and georeferencing technologies have opened new possibilities and opportunities to more accurately assess the evolution of land degradation and improvement and to determine their causes and associations with human welfare. For example, by overlaying georeferenced child mortality rates—an indicator of poverty—with change in vegetation—an indicator of land degradation or improvement—Bai et al. (2008b) showed a positive relationship between poverty and land degradation. They were also able to show the (surprising) negative relationship between population density and land degradation.

Institutions Responsible for Policy Actions Against LD Now Need to Evolve with the Current Scientific, Evidence-Based Knowledge of LD.

Bai et al. (2008b) also provided a radically different view of the location of land degradation. Whereas past global studies, which were largely based on expert opinion, tended to focus on arid and semiarid areas, leading to the notion that land degradation is a problem largely affecting dry areas, this study showed a negative relationship between aridity and land degradation. The authors showed that between 1981 and 2003, about 78 percent of the world's degraded land (measured in terms of loss of vegetation) is located in humid areas. These results have significant implications, in that the early focus on desertification in the 1970s had partly shaped the institutional setup and focus. The best example is the United Nations Convention to Combat Desertification (UNCCD), whose name reflects its focus on land degradation in dry areas. Although it is true that the impacts of land degradation in dry areas are severe and that land degradation affects some of the most vulnerable populations living in the most vulnerable environments, the extent and severity of land degradation in humid areas calls for more attention and action, at both the national and global levels. Institutional actors need to take note of the best currently available knowledge and science in setting policy programs targeting LD.

Despite the Technological Advances and GLADA's Findings, the GLADA Study has Weaknesses that Need to be Taken into Account in Future Studies.

The normalized difference vegetation index (NDVI) is a complex and abstract index used in the satellite observation of land degradation. The hypothesis is that variations in vegetation cover indicate either land degradation or improvement. Soils and their cover (for example, vegetation) enter the broad definition of land degradation, which means a decrease in vegetation cover is indeed a form of land degradation, and yet the NDVI does not convey any information about the state of soils. It is thus dangerous to interpret the NDVI as an indicator of land degradation in any other sense than a change in vegetation cover. Further, the NDVI fails to differentiate between forms of vegetation. For example, alien species encroachment could increase the NDVI and therefore be viewed as land improvement, when it is actually a form of land degradation. Vegetation is also determined by many other factors than land degradation or improvement. For example, although Bai et al. (2008b) showed a severe decrease of NDVI in Africa south of the equator, cereal productivity increased significantly in Cameroon, Malawi, and a few other countries. Such an increase was due to the use of improved crop varieties and land management practices. Atmospheric carbon fertilization has also increased NDVI, masking the actual land degradation (Vlek, Le, and Tamene 2010). The GLADA study also shows degradation in some areas that have sparse population density. For example, Gabon and Congo show the most severe land degradation; yet population density in the two countries is among the lowest in the region. From a socioeconomic perspective, the NDVI is an indicator that is dissociated from people and their social and economic relations: It assesses degradation in remote, unpopulated places equally to NDVI-based degradation that destroys livelihoods in other areas. For this reason, NDVI studies have so far had little policy impacts.

These and other shortcomings underscore the need to better calibrate the satellite data to address their shortcomings. For example, there is a need to establish sentinel study sites,[73] where in-depth analysis can be done to better understand and calibrate the relationship between remotely sensed biophysical and socioeconomic data with the actual land degradation or improvement. Additional data that cannot be collected using satellite imagery could also be collected from the sentinel sites—in particular, socioeconomic data to indicate the human relevance of the land degradation indicators. The results from the sentinel sites could then be extrapolated to the global level.

Understanding the Underlying Causes of land degradation will Help in the Design of Appropriate Actions for Preventing or Mitigating Land Degradation.

This review showed that understanding the underlying causes of land degradation is important for designing strategies for taking action to prevent or mitigate land degradation. The study also showed that the impact of one particular underlying cause of land degradation depends on the other underlying causes. For example, population density could lead to more severe land degradation if there are no strong institutions to regulate the behavior of communities or if market forces do not give land users an incentive to invest in land improvement. This situation suggests that taking action to prevent or mitigate land degradation requires the design of policies and strategies that will simultaneously address the multiple underlying causes of land degradation.

Of particular importance is the need to develop strong local institutions for land management at the community level and to provide incentives for individual land users to invest in land improvement. This task requires decentralization policies, which provide mandates and which facilitate the development of local institutions. Studies have shown that countries that have been investing in land improvement and providing incentives to land users have

73 Sentinel sites are selected for an in-depth study or data collection such that the sentinel sites are representative of a larger area or population. Results from the sentinel sites could then be interpolated to the larger area or population they represent.

seen greater improvement, despite their high population densities. For example, China has provided incentives for farmers in the western highlands to plant trees, and the Bai et al. (2008b) study showed a significant improvement of vegetation there. India's decentralized government also allows communities to form community-based watershed management (CBWM) committees. For instance, a switch from centrally managed watershed to CBWM in Tamil Nadu, India, in 2009) resulted in the water table receding and water availability increasing in the area. This significant change was largely due to the mandate given to local communities to manage and benefit from the watershed. Similar success stories have been observed by the International Forest Research Institute, which has been conducting research on community-based forest management in developing countries (Gibson, Williams, and Ostrom 2005).

Taking Action to Prevent Or Mitigate Land Degradation Requires an Economic Analysis of the Costs of Land Degradation and the Costs and Benefits of Preventing or Mitigating Land Degradation.

The economic analysis proposed in this study is the well-established concept of measuring the economic costs and benefits using total economic value of terrestrial ecosystem services, which comprise the on-site and off-site direct and indirect costs and benefits. Because both land degradation and action to prevent it have lagged effects, it is necessary to use dynamic modeling to determine the future costs and benefits, which requires long-term data collection and simulation analysis using well-calibrated models.

To analyze off-site costs and benefits requires an association of the benefits to the beneficiaries and of the costs to those who experience the negative impacts or implement the land conservation action. Such an analysis will, for instance, enable payment for ecosystem services (PES) schemes to enhance the adoption of land management practices that would otherwise not be profitable if land users were not compensated for their actions. Local, national, and international cooperation is required to ensure such collaboration.

Case studies from selected countries showed that the cost of preventing or mitigating land degradation is much less than the cost of land degradation. However, public investments in addressing land degradation in developing countries have been quite limited. One reason for such limited investment is the limited number of studies on the economics of land degradation. The global ELD assessment proposed in this study will close this gap at the global level. Furthermore, the proposed country-level studies are required to inform the national-level policymakers so they can take action to address land degradation.

To implement the global ELD assessment, a well-planned organizational arrangement is required to coordinate and harmonize resources, ensure mobilization and advocacy, and conduct the necessary research in multidisciplinary and transregional teams. A large number of institutions have already produced a number of studies in the past, and a global ELD initiative should take advantage of these by building on their strengths and using the data collected. For instance, a number of institutions have been collecting biophysical data on LD; however, collection of socioeconomic data is still limited. The new partnership should work hard to address the weaknesses of the current data, while nonetheless making use of the large amount of data freely shared by many publicly funded institutions.

Is this Partnership Possible, And What Should Be Done To Build A Global ELD Platform For Action?

The current increased awareness of land degradation and the growing interest in investing in land provide a great potential for mobilizing partnership around a global ELD assessment and, later, for implementing its recommendations. This would require champions of the cause to coordinate and facilitate action in both the policy and scientific spheres. It would also require experienced advocacy for mobilizing resources for a global ELD assessment and its implementation. We have proposed an institutional setup to that effect, in which all stakeholders of a global ELD initiative can meet and interact for the benefits of global action and investment against land degradation and their effects on human welfare. An open consultation process across all the different groups of the institutional setup would be a worthy initial phase and a continuation of the dialogue process of which this study was a part.

Appendix A: Supplementary Tables

Table A.1—Land degradation assessments on the national level

Author	Region	Methods Used
Meadows and Hoffman 2003	South Africa	Modeling of the potential impact of future climate changes on the nature and extent of land degradation in South Africa. The Climate System Model gave information on the interlinking of climate conditions and land degradation. Future climate change is a key challenge for developing economies of countries like South Africa
Sonneveld 2003	Ethiopia	Uses expert opinion to conduct a nationwide water erosion hazard assessment in Ethiopia.
Symeonakis and Drake 2004	Sub-Saharan Africa	The desertification monitoring system consists of four indicators: NDVI, rainfall use efficiency (RUE), surface runoff (using the Soil Conservation Service), and soil erosion (using a model parameterized by overland flow, vegetation cover, the digital soil maps, and a digital elevation model), calculated for 1996.
Klintenberg and Seely 2004	Namibia	Four primary indicators for Land Degradation Monitoring in Namibia: population pressure (population density), livestock pressure (distribution of boreholes and annual numbers of livestock), rainfall (index based on rainfall records), and erosion risk (based on gradient and soil characteristics for agroecological zones); 1971–1997
Foster 2006	Botswana	Using main methodologies for assessing land degradation in Botswana: Global Assessment of Human-Induced Land Degradation (GLASOD); remotely sensed images showing bush encroachment from 1997 and vegetation distribution in 1971 and 1994; agricultural productivity trends by region for 1980–1998; participatory studies from two degradation hot spots in Botswana; average annual rainfall for 1986–2000
Prince, Becker-Reshef, and Rishmawi 2009	Zimbabwe	Local Net Production Scaling[74]; information on land cover, precipitation (from the U.S. National Oceanic and Atmospheric Association, or NOAA), soil (Soil and Terrain Digital Database, or SOTER, and Zimbabwe soil map); net primary production (NPP) and Normalized Differenced Vegetation Index (NDVI) data from moderate resolution imaging spectroradiometer (MODIS)
Wessels et al. 2007	South Africa (and Limpopo)	Advanced very high resolution radiometer (AVHRR), NDVI, and modeled NPP were used to estimate vegetation production in South Africa. Human-induced signals were separated from natural land degradation by the use of RUE and residual trend (RESTREND).

Source: Author's compilation.

74 Local Net Production Scaling is the estimated potential production in homogeneous land capability classes. It models the actual productivity using remotely sensed observations.

Table A.2—Land degradation assessment on the local and subnational levels

Author	Region	Methods Used
Hill, Mégier, and Mehl 1995	Mediterranean ecosystems (test sites in the south of France and Greece)	Vegetation indexes Normalized Differenced Vegetation Index (NDVI), airborne imaging spectrometry, change detection with Landsat Thematic Mapper (TM) data, soil conditions
Hill et al. 1998	Greece, Crete	Long-term series of Landsat TM images (between 1984 and 1996), conversion into geographic information system (GIS) layers
Collado, Chuvieco, and Camarasa 2000	Argentina—San Luis Province	Comparing two Landsat images for 1982 (a more arid period) and 1992 (a humid period) of the area; difference picture of two Landsat images; Monitoring of the following: (i) enlargement of water bodies, (ii) increasing soil degradation because of increased grazing pressure, (iii) invasion of alien species while other palatable species disappear , (iv) displacement of sand dunes (using albedo monitoring[75]
Diouf and Lambin 2001	Senegal—Region Ferlo	Assess land cover modifications in Ferlo with rainfall data and rainfall use efficiency (RUE), advanced very high resolution radiometer (AVHRR), NDVI, data on soil types, changes in floristic composition, analysis of the resilience after drought
Gao, Zha, and Ni 2001	China—Yulin, Shaanxi provinces	Aerial photographs, one Landsat TM image and GIS; trend of desertification between 1960 and 1987 is modeled from changes in other land covers
Evans and Geerken 2004	Syria	Distinguish between climate- and human-induced dryland degradation based on evaluations of AVHRR, NDVI data, and rainfall data
Herrmann, Anyamba, and Tucker 2005	Africa—Sahel	Investigation of temporal and spatial patterns of vegetation greenness and rainfall variability and their interrelationship, based on NDVI time series for 1982–2003 and gridded satellite rainfall estimates.
Kiunsi and Meadows 2006	Tanzania—Monduli District, Northern Tanzania)	Three sets of land cover maps synchronized against long-term rainfall data (1960s, 1991, and 1999). The change detection, based on the land cover map set, gives information on changes in vegetation due to rainfall, which could be separated from changes in vegetation that occurred due to human impact.
Hein and de Ridder 2006	Africa—Sahel	Critical assessment of desertification by the use of RUE. Variability of RUE for the analysis of remote sensing imagery of semiarid rangelands with regard to natural and human-induced degradation of Sahelian vegetation cover.

75 Bare soils show a higher reflectance in the visible bands, while green vegetation strongly absorbs it. This method can be useful in assessing wind degradation by the movement of sand dunes.

Author	Region	Methods Used
Lu et al. 2007	Brazil—Western Brazilian Amazon	Mapping and monitoring of land use and land cover changes by the use of remote sensing (Landsat TM/ETM+ images). A surface cover index is developed to evaluate and map potential land degradation risks associated with deforestation and accompanying soil erosion in the rural settlements of the study area.
Prince et al. 2007	Africa—Sahel	Using RUE to describe the difficulty of estimating the RUE for nondegraded land at a regional scale. Answer to the article by Hein and de Ridder (2006).
Hill et al. 2008	European Mediterranean	Adaptation of the syndrome approach to the Iberian peninsula. Characterization of vegetation dynamics based on NDVI U.S. National Oceanic and Atmospheric Association (NOAA) AVHRR.
Helldén and Tottrup 2008	Mediterranean basin, Sahel from the Atlantic to the Red Sea, major parts of drylands in Southern Africa, China-Mongolia, and South America,	NOAA AVHRR data for desertification monitoring over a regional and global level; Global Inventory Modeling and Mapping Studies 8-kilometer global NDVI dataset; rainfall dataset for 1981–2003
Omuto and Vargas 2009	Somalia, Northwest	Risk of soil loss in northwestern Somalia; testing the use of pedometrics, remote sensing (Landsat ETM+ imagery), limited field data collection, and revised universal soil loss equation (RUSLE)
Gao and Liu 2010	China—Tongyu County, Northeast China	Monitoring land cover changes with satellite imagery. Change detection between data from 1992 and 2002.

Source: Own compilation.

Table A.3—Review of studies estimating off-site costs of land degradation (in chronological order)

Author	Country	Degradation process	Type of off-site cost	Off-site cost	Unit	Note
Clark 1985		Soil erosion	Total off-stream damage	1,100–3,100	US$ million	in 1980 dollars
			Total in-stream damage	2,100–10,000	US$ million	in 1980 dollars
Cruz et al. 1988	Philippines—Pantabangan Reservoir	Soil erosion	Reduction in service life of reservoir	1.11	Philippine pesos per hectare	
			Reduction in active storage and irrigation	12.99	Philippine pesos per hectare	
			Reduction in active storage and hydropower	2.91	Philippine pesos per hectare	
			Opportunity cost of dead storage for irrigation	575.55	Philippine pesos per hectare	
	Philippines—Magat Reservoir	Soil erosion	Reduction in service life of reservoir	0.1	Philippine pesos per hectare	
			Opportunity cost of dead storage for irrigation	365.61	Philippine pesos per hectare	
Magrath and Arens 1989	Indonesia, Java	Soil erosion	Irrigation system siltation	7.9–12.9	US$ million	
			Harbor dredging	1.4–3.5	US$ million	
			Reservoir sedimentation	16.3–74.9	US$ million	
Grohs 1994	Zimbabwe	Soil erosion	Sedimentation (productivity change approach)	0.6	Zimbabwean dollars	in 1989 dollars
			Sedimentation (replacement cost approach)	0.8–8.8	Zimbabwean dollars	In 1989 dollars
			Sedimentation (defensive expenditure)	1.0–12.5	Zimbabwean dollars	In 1989 dollars
Pimentel et al. 1995	United States	Water erosion	Recreational	2,440.0	US$ million per year	On- and off-site costs of erosion in United States: US$44 billion per year or $100/ha
			Water-storage facilities	841.8	US$ million per year	
			Navigation	683.2	US$ million per year	
			Other in-stream uses	1,09.08	US$ million per year	
			Flood damages	939.4	US$ million per year	

Author	Country	Degradation process	Type of off-site cost	Off-site cost	Unit	Note
			Water-conveyance facilities	244.0	US$ million per year	
			Water-treatment facilities	122.0	US$ million per year	
Pimentel et al. 1995	United States	Water erosion	Other off-stream uses	976.0	US$ million per year	
		Wind erosion	Exterior paint	18.5	US$ million per year	
			Landscaping	2,894.0	US$ million per year	
			Automobiles	134.6	US$ million per year	
			Interior, laundry	986.0	US$ million per year	
			Health	5,371.0	US$ million per year	
			Recreation	223.2	US$ million per year	
			Road maintenance	1.2	US$ million per year	
			Cost to business	3.5	US$ million per year	
			Cost to irrigation and conservation districts	0.1	US$ million per year	
Pretty 2000, 124	United Kingdom	Soil erosion	Damage to roads and property	4.00	Million pounds	Calculated for various off- and in-stream damages
			Traffic accidents	0.10	Million pounds	
			Footpath loss	1.19	Million pounds	
			Channel degradation	8.47	Million pounds	
Krausse et al. 2001	New Zealand	Soil erosion	Sedimentation	27.4	NZ-Mil. $	Calculated for various off- and in-stream damages; in 1998 dollars
Hansen et al. 2002, 211	United States	Erosion	Dredging	257	US$ million per year	In 1998 dollars; not included: sediment dredged by lake

Author	Country	Degradation process	Type of off-site cost	Off-site cost	Unit	Note
						or ocean action
Vieth, Gunatilake, and Cox 2001, 145	Sri Lanka – Upper Mahareli Watershed	Soil erosion	Reduction in irrigated area	0.080	US-Mil. $	In 1993 dollars
			Reduction in hydropower production	0.288	Mil. $	In 1993 dollars
Vieth, Gunatilake, and Cox 2001, 145	Sri Lanka – Upper Mahareli Watershed	Soil erosion	Cost of water purification	0.080	Mil. $	In 1993 dollars
Tegtmeier and Duffy 2004, 4	United States	Soil erosion	Cost to water industry	277.0–831.1	US$ million	In 2002 dollars
			Cost to replace lost capacity of reservoirs	241.8–6,044.5	US$ million	In 2002 dollars
			Water-conveyance costs	268.0–790.0	US$ million	In 2002 dollars
			Flood damages	190.0–548.8	US$ million	In 2002 dollars
			Damage to recreational activities	540.1–3,183.7	US$ million	In 2002 dollars
			Cost to navigation (shipping damage, dredging)	304.0–338.6	US$ million	In 2002 dollars
			In-stream impacts (fisheries, preservation value)	242.2–1,218.3	US$ million	In 2002 dollars
			Off-stream impacts (industrial uses, steam power plants)	197.6–439.7	US$ million	In 2002 dollars
Colombo et al. 2005	Spain—Andalusian region	Soil erosion	Landscape desertification: small/medium improvement	17.428–22.88	Euro	Implicit price
			Surface and groundwater quality: medium/high quality	21.865–29.352	Euro	

Author	Country	Degradation process	Type of off-site cost	Off-site cost	Unit	Note
			Flora and fauna quality: improvement to medium/good quality	14.992–17.765	Euro	
			Jobs created (number)	0.102	Euro	
Hansen and Hellerstein 2007	United States	Soil erosion	Marginal benefit of a 1-ton reduction in soil erosion	0–1.38	US$	
Nkonya et al. 2008b	Kenya	Soil erosion, loss of vegetation	Carbon sequestration, siltation (cost of treatment and purification of water)	24	KES per hectare	
Richards 1997, 24	Bolivia—Taquina watershed	Soil erosion	Flood prevention	2.30	US-Mil. $	Annual benefit from year 7–50 after installment of conservation measures
			Aquifer recharge	7.80	US-Mil. $	Annual benefit from years 7–20 after installment of conservation measures
Feather, Hellerstein, and Hansen 1999	United States	Soil erosion	Recreation	80	US$ million	Benefit
			Wildlife viewing	348	US$ million	Benefit
			Hunting	36	US$ million	Benefit

Source: Author's compilation.

Table A.4—Review of studies estimating the net present value (NPV) of returns to different conservation measures (in chronological order)

Author	Country	Area	Conservation measure	Crop	Discount rate	Time horizon (years)	NPV	Unit	Note	Soil erosion with conservation	Soil erosion without conservation
Lutz, Pagiola, and Reiche 1994	Costa Rica	Barva	Diversion ditches	Coffee	0.2		–920	US$	Net gains		
		Tierra Blanca	Diversion ditches	Potato	0.2		–3,440	US$	Net gains		
		Turrubares	Diversion ditches	Coco yam	0.2		1,110	US$	Net gain		
		Turrubares	Terraces	Coco yam	0.2		4,140	US$	Net gains		
	Dominican Republic	El Naranjal	Diversion ditches	Pigeon peas, peanuts, beans	0.2		–132	US$	Net gains		
	Guatemala	Patzité	Terraces	Corn	0.2		–156	US$	Net gains		
	Haiti	Maissade	Ramp pay	Corn, sorghum	0.2		1,180	US$			
		Maissade	Rock walls	Corn, sorghum	0.2		956	US$	Net gains		
	Honduras	Tatumbla	Diversion ditches	Corn	0.2		909	US$	Net gains		
	Honduras	Yorito	Diversion ditches	Corn	0.2		83	US$	Net gains		
	Panama	Coclé	Terraces	Rice, corn, yucca beans	0.2		34	US$	Net gains		

Author	Country	Area	Conservation measure	Crop	Discount rate	Time horizon (years)	NPV	Unit	Note	Soil erosion with conservation	Soil erosion without conservation
Partap and Watson 1994	Philippines		Hedgerow	Corn	0.10	6	61	US$ per hectare			
					0.05	6	230	US$ per hectare			
Bishop and Allen 1989	Mali	Nationwide			0.1	10	31	US$ million	Impact of soil loss on yield (beta) = 0.004		
Pagiola 1996	Kenya	Kitiu District	Terraces	Maize–beans	0.1	50	Distribution of revenues over time		Net gain depending on slope (5%, 10%, 15%, 20%)		
Nelson et al. 1998	Philippines		Hedgerow intercropping	Maize	0.25	10	Distribution of NPV over time		NPV for open field, fallow, and hedgerows	1 ton per hectare per year (ton/ha/yr)	190 ton/ha/yr
					0.2	20	Distribution of NPV over time		NPV for open field, fallow, and hedgerows	1 ton/ha/yr	190 ton/ha/yr
Gebremedhin, Swinton, and Tilahun 1999	Ethiopia	Tigray	Stone terraces	Wheat–wheat–fava beans rotation	0.15	30	277	US$ per acre	Net gain; considers yield differences		
Shively 1999	Philippines	Barangay Bansalam	Hedgerows	Corn	0.10	10	34	US$ per hectare	Net gain		

Author	Country	Area	Conservation measure	Crop	Discount rate	Time horizon (years)	NPV	Unit	Note	Soil erosion with conserva tion	Soil erosion without conserv ation
		in Davao del Sur Province									
				Corn	0.05	10	92	US$ per hectare	Net gain		
					0.50	30	1	US$ per hectare	Net gain		
Shiferaw and Holden 2001	Ethiopia	Andit Tid	Level bund	Barley	0		62,743	Birr per hectare (Birr/ha)	Also available for Anjeni area and other crops (wheat, fava beans); net gains, 16% area loss due to conservation	10 ton/ha/yr	42 ton/ha/yr
Shiferaw and Holden 2001	Ethiopia				0.05		–2,409	Birr/ha		10 ton/ha/yr	42 ton/ha/yr
					0.10		–2,306	Birr/ha		10 ton/ha/yr	42 ton/ha/yr
					0.20		–1,561	Birr/ha		10 ton/ha/yr	42 ton/ha/yr
			Graded bund	Barley	0		–398	Birr/ha		25 ton/ha/yr	42 ton/ha/yr
					0.05		–3,888	Birr/ha		25 ton/ha/yr	42 ton/ha/yr
					0.10		–2,695	Birr/ha		25 ton/ha/yr	42 ton/ha/yr
					0.20		–1,668	Birr/ha		25 ton/ha/yr	42 ton/ha/yr

Author	Country	Area	Conservation measure	Crop	Discount rate	Time horizon (years)	NPV	Unit	Note	Soil erosion with conservation	Soil erosion without conservation
			Level Fanya juu	Barley	0		–44,531	Birr/ha		12 ton/ha/yr	42 ton/ha/yr
					0.05		–3,248	Birr/ha		12 ton/ha/yr	42 ton/ha/yr
					0.1		–2,969	Birr/ha		12 ton/ha/yr	42 ton/ha/yr
					0.20		–2,136	Birr/ha		12 ton/ha/yr	42 ton/ha/yr
			Graded fanya juu	Barley	0		9,453	Birr/ha		20 ton/ha/yr	42 ton/ha/yr
					0.05		–4,039	Birr/ha		20 ton/ha/yr	42 ton/ha/yr
					0.10		–3,176	Birr/ha		20 ton/ha/yr	42 ton/ha/yr
					0.20		–2,192	Birr/ha		20 ton/ha/yr	42 ton/ha/yr
			Grass tripes	Barley	0		41,027	Birr/ha		15 ton/ha/yr	42 ton/ha/yr
					0.05		–1,259	Birr/ha		15 ton/ha/yr	42 ton/ha/yr
Shiferaw and Holden 2001	Ethiopia				0.10		–1,432	Birr/ha		15 ton/ha/yr	42 ton/ha/yr
					0.20		–937	Birr/ha		15 ton/ha/yr	42 ton/ha/yr

Author	Country	Area	Conservation measure	Crop	Discount rate	Time horizon (years)	NPV	Unit	Note	Soil erosion with conservation	Soil erosion without conservation
Posthumus and de Graaf 2005	Peru	Pacucha, field 1	Terraces		0.1	10	187	Neue Sol S/.	Net gain		
		Field 2					75	Neue Sol S/.	Net gain		
		Field 3					–96	Neue Sol S/.	Net gain		
		Field 4					–1,731	Neue Sol S/.	Net gain		
		Field 5					–869	Neue Sol S/.	Net gain		
		Field 6					–1,331	Neue Sol S/.	Net gain		
		Field 7					–2,344	Neue Sol S/.	Net gain		
		Field 8					–707	Neue Sol S/.	Net gain		
		Field 9					603	Neue Sol S/.	Net gain		
		Field 10					–906	Neue Sol S/.	Net gain		
		Field 11					–1,122	Neue Sol S/.	Net gain		
Nkonya et al. 2008b	Kenya		Agroforestry plus organic and inorganic fertilizer, fanya juu, fanya chini	Maize	0.1	50	152.00	In 1,000 KES per hectare	Net gain, private NPV		

Author	Country	Area	Conservation measure	Crop	Discount rate	Time horizon (years)	NPV	Unit	Note	Soil erosion with conservation	Soil erosion without conservation
			Agroforestry plus organic and inorganic fertilizer, fanya juu, fanya chini	Maize	0.1	50	176.05	In 1,000 KES per hectare	Net gain, social NPV (plus off-site effects)		
World Bank 2009	Niger		Tree plantations		0.1		307,000	FCFA/ha	Assumption: Low fodder value		
					0.1		125,000	FCFA/ha	High fodder value		
			Protected Areas		0.1		239,000	FCFA/ha	High fodder value		
					0.1		118,000	FCFA/ha	Low fodder value		
			Soil and water conservation				Not profitable				

Source: Author's compilation.

Notes: Ramp pay = crop stubble laid out along the contour, supported by stakes, and covered with soil; Fanya juu = throwing soil uphill; Fanya chini = throwing soil downhill.

Table A.5—Costs of land degradation (mainly soil erosion)

Author	Country	Costs	Unit	% of GDP	% of agricultural GDP	Note
Dregne and Chou 1992	World	42	US$ billion			
Huang and Rozelle 1995	China	700	US$ million		< 1%	
Solorzano et al. 1991	Costa Rica				5–13% of annual value added in agriculture	
FAO 1986	Ethiopia				< 1%	
Sutcliffe 1993	Ethiopia	155	US$ million		5%	
Bojö and Cassells 1995	Ethiopia	130	US$ million		3%	
Sonneveld 2002	Ethiopia				2.93%	
Convery and Tutu 1990	Ghana	166.4	US$ million		5%	
Diao and Sarpong 2007	Ghana	4.2	US$ billion (2006–2015)	18	5%	SLM-practices would generate an aggregate economic benefit of US$6.4 billion over the 2006–2015 period.
Magrath and Arens 1989	Indonesi, Java	340–406	US$ million	GDP growth per year	3% (Berry, Olson, and Campbell 2003)	
Cohen, Brown, and Shephard 2006	Kenya			3.80%		
Bojö 1991	Lesotho	0.3	US$ million		< 1%	
Eaton 1996	Malawi				3%	
Bishop and Allen 1989	Mali	2.9–11.6	US$ million		< 1%	
van der Pol 1992	Mali	59	US$ million			
McIntire 1994	Mexico				2.7–12.3%	10% discount rate
McKenzie 1994	South Africa				4%	

Norse and Saigal 1992	Zimbabwe	99.5	US$ million		8%	
Grohs 1994	Zimbabwe	0.7–2.1	US$ million	< 1%	0.36%	In 1988/1989 dollars
Stocking 1986	Zimbabwe	117	US$ million		9%	In 1986 dollars
Berry et al. 2003	China			4%		
	Ethiopia	139	US$ million	4%	0.2–0.5%	Direct effects
	Mexico	3.2	US$ billion			
Berry et al. 2003	Rwanda	23	US$ million		3.50%	Direct effects
Bishop 1995	Mali	1.1–7.3	US$ million	1.51%	3.38% (3–13% in Yesuf et al. 2005)	beta=0,004, beta-factor: sensitivity to soil erosion, values for different betas calculated
	Malawi	13	Mil. US-$		2,4% (17-55% in Yesuf et al., 2005)	Beta = 0.004
Young 1993	South and Southeast Asia				7%	
	India				5%	
	Pakistan				5%	
Drechsel and Gylele 1999	Mali				5.5–6.5%	
	Madagascar				6–9%	
	Malawi				9.5–11%	
	Ghana				4–5%	
	Ethiopia				10–11%	

Source: Author's compilation.

Appendix B: Supplementary Figures

Figure B.1—Land use systems of the world

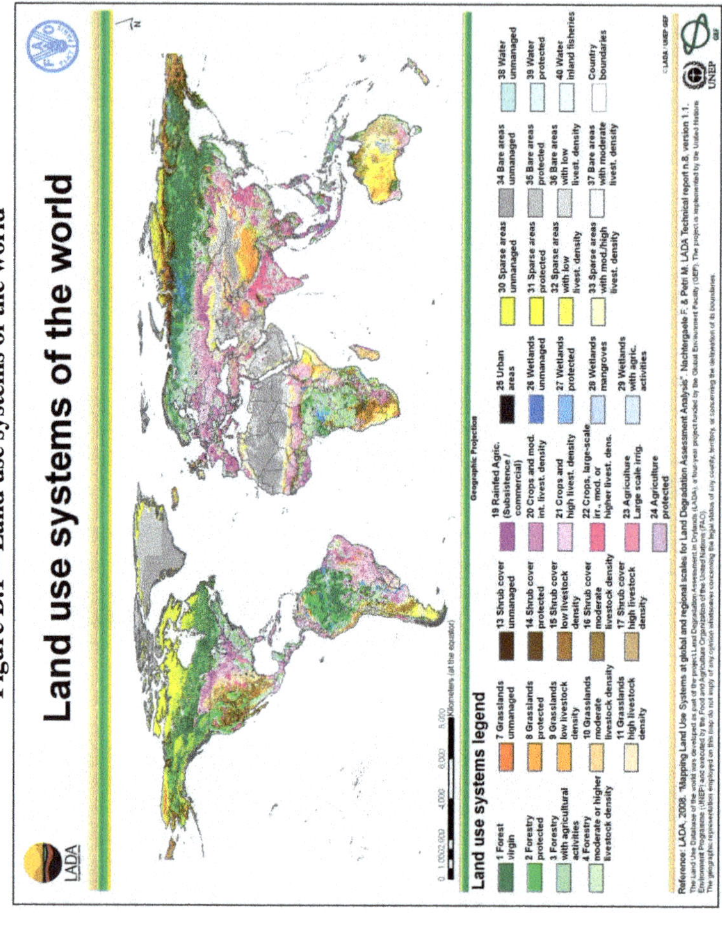

Source: LADA, 2008

Figure B.2—GLADA output

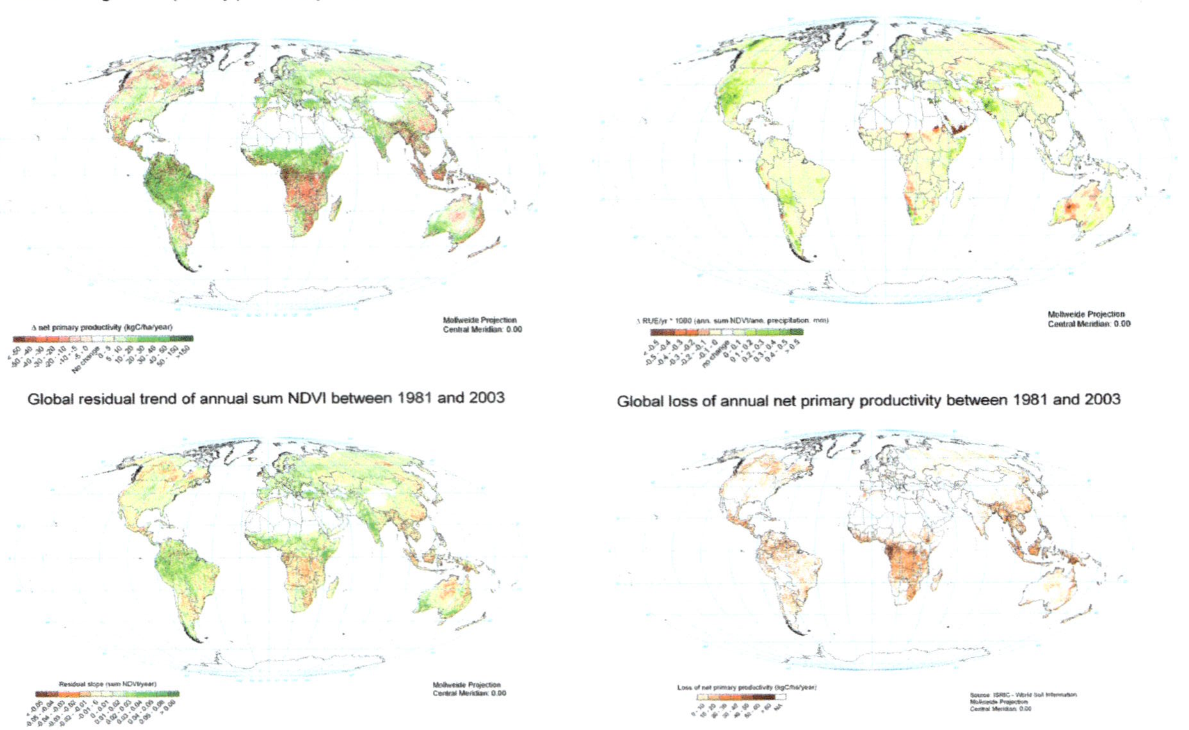

Source: FAO, available at: www.fao.org/geonetwork/srv/en/main.search?any=glada.

Figure B.3—Biophysical Status Index (BSI), GLADIS

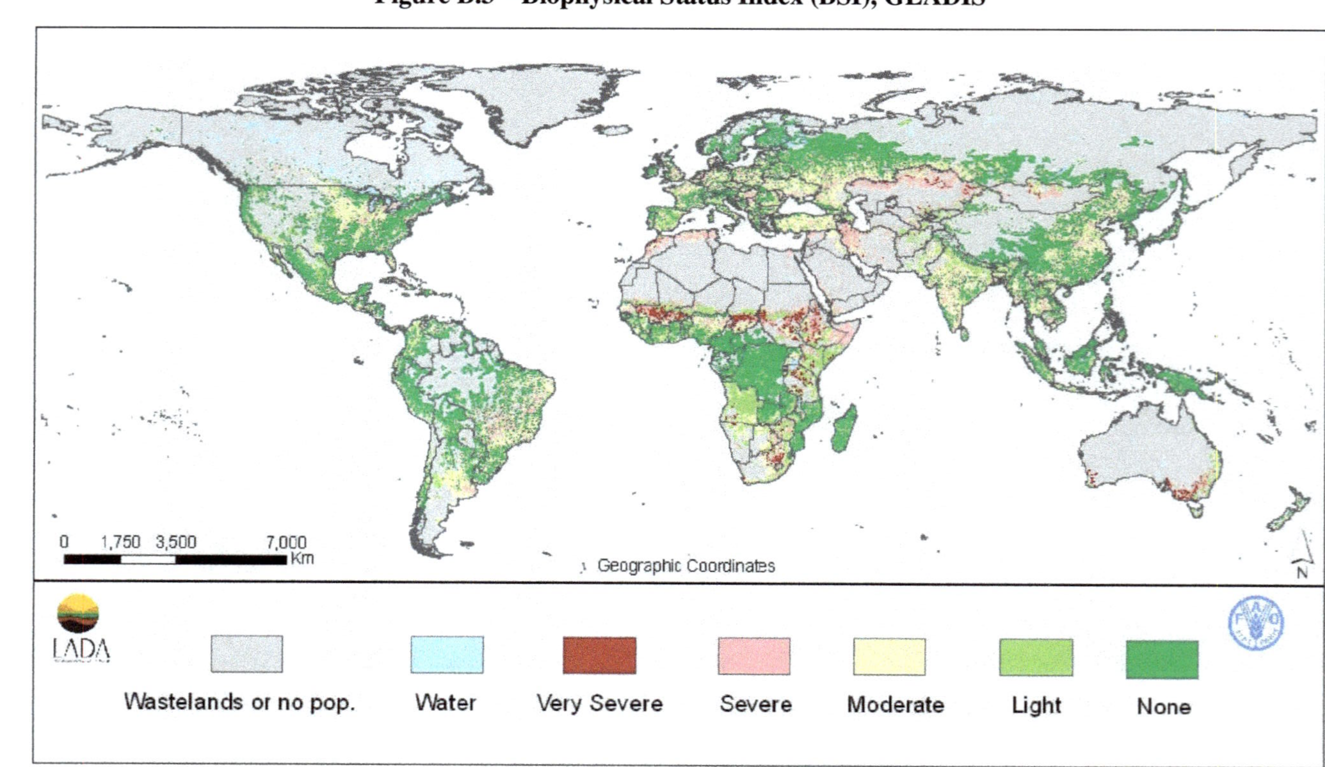

Source: Nachtergaele et al. 2010

Figure B.4—Goods and services severely affected, GLADIS

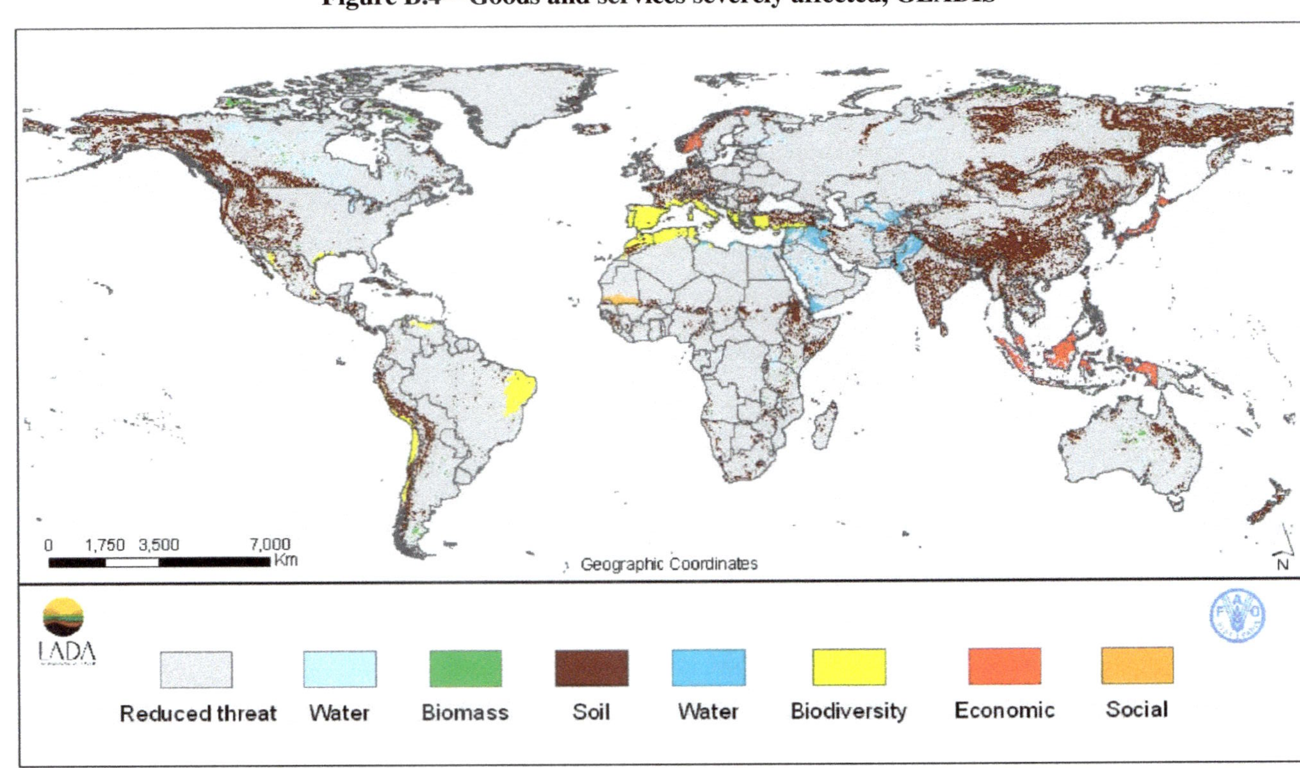

Source: Nachtergaele et al. 2010.

Abbreviations and Acronyms

ADB	Asian Development Bank
AI	Aridity Index
APSIM	Agricultural Production Systems Simulator
ASSOD	Assessment of Soil Degradation in Asia and Southeast Asia
AVHRR	Advanced Very High Resolution Radiometer
BDI	Biophysical Degradation Index
BMZ	German Ministry for Economic Cooperation and Development
BSI	Biophysical Status Index
CBA	Cost–Benefit Analysis
CBD	Convention on Biological Diversity
CBWM	Community-Based Watershed Management
CDM	Clean Development Mechanism
CER	Certified Emission Reduction
CERES	Crop Environment Resource Synthesis
CGE	Computable General Equilibrium Model
CIESIN	Center for International Earth Science Information Network
CMEPSP	Commission on the Measurement of Economic Performance and Social Progress
COP	Conference of Parties
LD	Land Degradation
DDP	Dryland Development Paradigm
EC_e	Electrical Conductivity of a Saturated Soil Extract
ELD	Economics of Land Degradation
EPIC	Erosion Productivity Impact Calculator
ESRI	Environmental Systems Research Institute
ESSI	Ecosystem Service Status Index
FAO	Food and Agriculture Organization (UN)
FAOSTAT	Food and Agriculture Organization Statistic
FMNR	Farmer-Managed Natural Regeneration
GAEZ	Global Agro-Ecological Zone
GCM	General Circulation Model
GDP	Gross Domestic Product
GEF	Global Environment Facility
GIMMS	Global Inventory Modeling and Mapping Studies

GIS	Geographical Information System
GLADA	Global Land Degradation Assessment
GLADIS	Global Land Degradation Information System
GLASOD	Global Assessment of Human-Induced Land Degradation
GRUMP	Global Rural–Urban Mapping Project
GTOS	Global Terrestrial Observing Systems
HANPP	Human Appropriation of Net Primary Production
HDI	Human Development Index
HEP	Hydroelectric Power
IFPRI	International Food Policy Research Institute
IFRI	International Forestry Resources and Institutions
IMR	Infant Mortality Rate
IPCC	Intergovernmental Panel on Climate Change
IRR	Internal Rate of Return
ISFM	Integrated Soil Fertility Management
ISRIC	International Soil Reference and Information Center
ISSS	International Society of Soil Science
LADA	Land Degradation Assessment in Drylands
LDI	Land Degradation Index
LDII	Land Degradation Impact Index
LECZ	Low-Elevation Coastal Zone
MA	Millennium Ecosystem Assessment
MAC	Marginal Abatement Cost Curve
MEA	Multilateral Environmental Agreement
MODIS	Moderate Resolution Imaging Spectroradiometer
MRR	Marginal Rates Of Return
MSA	Mean Species Abundance
NAP	National Action Plan
NAPA	National Adaptation Program of Action
NASA	National Aeronautics and Space Administration
NBSAP	National Biodiversity Strategy and Action Plan
NDVI	Normalized Differenced Vegetation Index
NGO	Nongovernmental Organization
NOAA	U.S. National Oceanic and Atmospheric Association
NPP	Net Primary Production

NPV	Net Present Value
NRCS	Natural Resources Conservation Service
NRM	Natural Resource Management
NSC	National Steering Committee
NTFP	Nontimber Forest Product
ODA	Overseas Development Administration
OECD	Organization for Economic Cooperation and Development
OLS	Ordinary Least Regression
PERFECT	Productivity, Erosion and Runoff Functions, to Evaluate Conservation Techniques
PES	Payment for Ecosystem Services
RADAR	Radio Detection and Ranging
REDD	Reducing Emissions from Deforestation and Forest Degradation
RESTREND	Residual Trend
RUE	Rainfall Use Efficiency
RUSLE	Revised Universal Soil Loss Equation
SAVI	Soil-Adjusted Vegetation Index
SCAR	Soil Conservation in Agricultural Regions
SEDAC	Socioeconomic Data and Application Center
SLEMSA	Soil Loss Estimation Model for Southern Africa
SLM	Sustainable Land Management
SLWM	Sustainable Land And Water Management
SOTER	Soil and Terrain Digital Database
SWC	Soil and Water Conservation
TEEB	The Economics of Ecosystems and Biodiversity
TEV	Total Economic Value
TFP	Total Factor Productivity
TLU	Tropical Livestock Unit
TM	Thematic Mapper (Landsat)
UNCCD	United Nations Convention to Combat Desertification
UNCED	United Nations Conference on Environment and Development
UNCOD	United Nations Conference on Desertification
UNEP	United Nations Environment Program
UNESCO	United Nations Educational, Scientific, and Cultural Organization

UNFCCC	United Nations Framework Convention on Climate Change
UN-REDD	Reducing Emissions from Deforestation and Forest Degradation
USDA	United States Department of Agriculture
USLE	Universal Soil Loss Equation
WAD	World Atlas of Desertification
WEPP	Water Erosion Prediction Model
WMO	World Meteorological Organization
WOCAT	World Overview of Conservation Approaches and Technologies

References

Abelson, P. 1979. Cost Benefit Analysis and Environmental Problems. Saxon House, London, UK.

Adam, T., C. Reij, T. Abdoulaye, M. Larwanou, and G. Tappan. 2006. *Impacts des Investissements dans la Gestion des Resources Naturalles (GRN) au Niger: Rapport de Synthese.* Niamey, Niger: Centre Régional d'Enseignement Specialisé en Agriculture (CRESA). Niamey, Niger et l'Université libre d'Amsterdam, Pays-Bas.

Adegbidi, A., E. Gandonou, and R. Oostendorp. 2004. "Measuring the Productivity from Indigenous Soil and Water Conservation Technologies with Household Fixed Effects: A Case Study of Hilly Mountainous Areas of Benin." *Economic Development and Cultural Change* 52 (2): 313–346.

Aglietta, M. 2010. "Sustainable Growth: Do We Really Measure the Challenge?" Paper presented at the Eighth Agence Française de Développement/European Development Research Network Conference, Paris, December 1.

Agrawal, A., and J. Ribot. 1999. "Accountability in Decentralization: A Framework with South Asian and West African Cases." *The Journal of Developing Areas* 33 (4): 473–502.

Ahmad M. 2000. Water Pricing and Markets in the Near East: Policy Issues and Options. *Water Policy* 2(3): 229-242.

Akhtar-Schuster, M., H. Bigas, and R. Thomas. 2010. *Monitoring and Assessment of Desertification, Land Degradation, and Drought: Knowledge Management, Institutions, and Economics.* White Paper of the DSD Working Group 3. Association of DesertNet International United Nations University – Institute for Water, Environment and Health Dryland Science for Development Consortium. February 5, 2010.

Alfsen, K. H., M. A. de Franco, S. Glomsrod, and T. Johnson. 1996. "The Cost of Soil Erosion in Nicaragua." *Ecological Economics* 16 (2): 129–145.

Alston, L.J., G.D Libecap and R. Schneider. 1995. "Property Rights and the Preconditions for Markets: The Case of the Amazon Frontier". *J . Institut. Theoret. Econom.* 151: 89-107.

Amsalu, A., and J. de Graaf. 2007. "Determinants of Adoption and Continued Use of Stone Terraces for Soil and Water Conservation in an Ethiopian Highland." *Ecological Economics* 61: 294–302.

Anderson, K., ed. 2009. Distortions to Agricultural Incentives: A Global Perspective 1955–2007. Washington, DC: World Bank.

Anderson, K., and E. Ostrom. 2008. "Analyzing Decentralized Resource Regimes from a Polycentric Perspective." *Policy Science* 41: 71–93.

Anlay, Y., A. Bogale, and A. Haile-Gabriel. 2007. "Adoption Decision and Use Intensity of Soil and Water." *Land Degradation and Development* 18 (3): 289–302.

Ariga, J., T. Jayne, and J. Nyoro. 2006. Factors Driving the Growth in Fertilizer Consumption in Kenya, 1990–2005: Sustaining the Momentum in Kenya and Lessons for Broader Replicability in Sub-Saharan Africa. Food Security Collaborative Working Papers 55167. East Lansing: Michigan State University, Department of Agricultural, Food, and Resource Economics. http://ideas.repec.org/s/ags/midcwp.html.

Asiema, J. K. and F. D. Situma. 1994. Indigenous Peoples and the Environment: The Case of the Pastorial Maasai of Kenya;. *Colorado Journal of International environmental law and policy* 5: 149-171.

Atwood, D. A. 1990. "Land Registration in Africa. The Impact on Agricultural Production." *World Development* 18 (5): 659–671.

Aune, J., and R. Lal. 1995. "The Tropical Soil Productivity Calculator: A Model for Assessing Effects of Soil Management on Productivity." In *Soil Management: Experimental Basis for Sustainability and Environmental Quality* edited by R. Lal and B. Stewart, 499–520. Lewis Publishers, Coca Raton, Florida, USA.

Bai, Z., and D. Dent. 2008. *Land Degradation and Improvement in Argentina. 1. Identification by Remote Sensing*. Wageningen, The Netherlands: International Soil Reference Information Center—World Soil Information.

Bai, Z. G., D. L. Dent, L. Olsson, and M. E. Schaepman. 2008a. "Proxy Global Assessment of Land Degradation." *Soil Use and Management* 24 (3): 223–234.

———. 2008b. Global Assessment of Land Degradation and Improvement. 1. Identification by Remote Sensing. GLADA Report 5 (November). Wageningen, The Netherlands.

Balmford, A., A. Bruner, P. Cooper, R. Costanza, S. Farber, R. E. Green, M. Jenkins, P. Jefferiss, V. Jessamy, J. Madden, K. Munro, N. Myers, S. Naeem, J. Paavola, M. Rayment, S. Rosendo, J. Roughgarden, K. Trumper, and R. K. Turner. 2002. "Economic Reasons for Conserving Wild Nature." *Science* 297: 950-953.

Balmford, A., A. Rodrigues, M. Walpole, P. ten Brink, M. Kettunen, and L. Braat. 2008. *The Economics of Biodiversity and Ecosystems: Scoping the*

Science. Final Report. Cambridge, UK: European Commission (contract: ENV/070307/2007/486089/ETU/B2).

Barbier E. B. 1990. The Farm-Level Economics of Soil Conservation: The Uplands of Java. *Land Economics* 66(2):199-211.

———. 1996. "The Economics of Soil Erosion: Theory, Methodology, and Examples." Paper based on a presentation to the Fifth Biannual Workshop on Economy and Environment in Southeast Asia, Singapore, November 28–30.

———. 1997. "The Economic Determinants of Land Degradation in Developing Countries." *Philosophical Transactions of the Royal Society of London B* (352): S891–899.

———. 1998. "The Economics of Environment and Development." In *The Economics of Environment and Development: Selected Essays,* edited by E. Barbier, 281–307. Cheltenham, UK: Edward Elgar.

———. 1999. "The Economics of Land Degradation and Rural Poverty Linkages in Africa." Proceedings of the United Nations University/Institute for Natural Resources in Africa, Annual Lectures on Natural Resource Conservation and Management in Africa. Tokyo: United Nations University Press.

———. 2000. "The Economic Linkages between Rural Poverty and Land Degradation: Some Evidence from Africa." *Agriculture, Ecosystem, and Environment* (82): 355–370.

Barbier, E., A. Markandya, and D. W. Pearce. 1998. "Environmental Sustainability and Cost–Benefit Analysis." In *The Economics of Environment and Development: Selected Essays,* edited by E. Barbier, S54–64. Cheltenham, UK: Edward Elgar.

Barrett, S. 1991. "Optimal soil conservation and the reform of agricultural pricing policies." *Journal of Development Economics* 36: 167- 187.

Baret F., O. Hagolle, B. Geiger, P. Bicheron, B. Miras, M. Huc, B. Berthelot, et al. 2007. "LAI, fAPAR and fCover CYCLOPES Global Products Derived from Vegetation Part 1: Principles of the Algorithm." *Remote Sensing of Environment* 110 (3): 275–286.

Barro, R. J. 1999. "Determinants of Democracy." *Journal of Political Economy.* 107(6):158–183.

Barrow, C. J. 1991. Land Degradation: Development and Breakdown of Terrestrial Environments. Cambridge, UK: Cambridge University Press.

Beintema, N., and G. Stads. 2004. "Sub-Saharan African Agricultural Research: Recent Investment Trends." *Outlook on Agriculture* 33 (4): 239–246.

Bejranonda, S., F. J. Hitzhusen, and D. Hite. 1999. "Agricultural Sedimentation Impacts on Lakeside Property Values." *Agricultural and Resource Economics Review* 28 (2): 208–218.

Bekele, W., and L. Drake. 2003. "Soil and Water Conservation Decision Behavior of Subsistence Farmers in the Eastern Highlands of Ethiopia: A Case Study of the Hunde-Lafto Area." *Ecological Economics* 46: 437–451.

Benin S., E. Nkonya, G. Okecho, J. Pender, S. Nahdy, S. Mugarura, E. Kato, and G. Kayobyo. 2007. *Assessing the Impact of the National Agricultural Advisory Services (NAADS) in the Uganda Rural Livelihoods.* IFPRI Discussion Paper 00724. Washington, DC: International Food Policy Research Institute.

Berkes, F. 2002. "Cross-Scale Institutional Linkage Perspective from the Bottom Up." In *The Dram of the Commons: Committee on the Human Dimensions of Global Change,* edited by O. E. T. Dietz, N. Polsak, P. Stern, S. Stovich, and E. Weber. Washington, DC: Science Academy.

———. 2007. "Community-Based Conservation in a Globalized World." *Proceedings of the National Academy of Sciences of the United States of America* 104 (39): 15188–15193.

Berry, L., J. Olson, and D. Campbell. 2003. "Assessing the Extent, Cost, and Impact of Land Degradation at the National Level." Findings from seven pilot case studies. Commissioned by Global Mechanism, with support from the World Bank.

Berry, S. 1993. No Condition Is Permanent: The Social Dynamics of Agrarian Change in Sub-Saharan Africa. Madison: University of Wisconsin Press.

Besley, T. 1995. "Property Rights and Investment Incentives: Theory and Evidence from Ghana." *The Journal of Political Economy* 103 (5): 903–937.

Bhutta M. N., L.K. Smedema. 2007. One Hundred Years of Waterlogging and Salinity Control in the Indus Valley, Pakistan: A Historical Review. *Irrigation and Drainage* 56(1):S81–S90.

Birendra, K. C., and S. Nagata. 2006. "Refugee Impact on Collective Management of Forest Resources: A Case Study of Bhutanese Refugees in Nepal's Eastern Terai Region." *Journal of Forest Research* 11: 305–311.

Bishop, J. 1995. The Economics of Soil Degradation: An Illustration of the Change in Productivity Approach to Valuation in Mali and Malawi. International Institute for Environment and Development, LEEC Discussion Paper 95-02. London: IIED.

Bishop, J., and J . Allen. 1989. *The On-Site Cost of Soil Erosion in Mali.* Environment Working Paper 21. Washington DC: The World Bank.

Byiringiro F., and T. Reardon. 1996. "Farm productivity in Rwanda: Effects of Farm Size, Erosion, and Soil Conservation Investments." *Agricultural Econonomics* 15(2): 127–136.

Blaikie, P., and H. Brookfield. 1987. *Land Degradation and Society.* New York: Richard Clay.

Blench R., 1999. Extensive Pastoral Livestock Systems: Issues and Options for the Future. FAO. Rome.

Bojö, J.. 1991. "The Economics of Land Degradation: Theory and Applications to Lesotho." The Stockholm School of Economics, Stockholm.

Bojö, J. 1996. "The Costs of Land Degradation in Sub-Saharan Africa." *Ecological Economics* 16: 161–173.

Bojö, J. and D. Cassells. 1995. "Land Degradation and Rehabilitation in Ethiopia. A Reassessment." World Bank. Washington DC.

Borlaug, N. E. 2000. *The Green Revolution Revisited and the Road Ahead. Special 30th Anniversary Lecture.* Oslo: Norwegian Nobel Institute.

Börner, J. 2006. "A Bio-Economic Model of Small-Scale Farmers' Land Use Decisions and Technology Choice in the Eastern Brazilian Amazon." PhD Dissertation, Zentrum für Entwicklungsforschung. Bonn: ZEF.

Boserup, E. 1965. The Conditions of Agricultural Growth: The Economics of Agrarian Change under Population Pressure. New York: Aldine Press.

———. 1981. *Population and Technology.* Oxford: Blackwell.

Bossio, D., K. Geheb, and W. Critchley. 2010. "Managing Water by Managing Land: Addressing Land Degradation to Improve Water Productivity and Rural Livelihoods." *Agricultural Water Management* 97: 536–542.

Bot, A., F. O. Nachtergaele, and A. Young. 2000. Land Resource Potential and Constraints at Regional and Country Levels. Food and Agriculture Organization of the United Nations (FAO), Rome, Italy.

Boumans J. H., J. W. van Hoorn, G. P. Kruseman, and B. S. Tanwar. 1988. "Water Table Control, Reuse, and Disposal of Drainage Water in Haryana." *Agricultural Water Management* 14 (1–4): 537–545. Boyd, C., and T. Slaymaker. 2000. "Re-examining the 'More People, Less Erosion'

Hypothesis: Special Case or Wider Trend?" *Natural Resource Perspective* 63: 1–6.

Boyd, J., and S. Banzhaf. 2007. "What Are Ecosystem Services? The Need for Standardized Environmental Accounting Units." *Ecological Economics* 63: 616–626.

Brasselle, F., A. Brasselle, F. Gaspart, and J.-P. Platteau. 2002. "Land Tenure Security and Investment Incentives: Puzzling Evidence from Burkina Faso." *Journal of Development Economics* 67: 373–418.

Bravo-Ureta, B. E., D. Solis, H. Cocchi, and R. E. Quiroga. 2006. "The Impact of Soil Conservation and Output Diversification on Farm Income in Central American Hillside Farming." *Agricultural Economics* 35: 267–276.

Bruyninckx, H. 2004. "The Convention to Combat Desertification and the Role of Innovative Policy-Making Discourses: The Case of Burkina Faso." *Global Environmental Politics* 4 (3): 107–127.

Buenemann, M., C. Martius, J. Jones, S. Hermann, D. Klein, M. Mulligan, M. S. Reed, et al. (2011). "Integrative Geospatial Approaches for the Comprehensive Monitoring and Assessment of Land Management Sustainability: Rationale, Potentials, and Characteristics." *Land Degradation and Development,* 22: 226–239.

Bunning, S., and D. Ndiaye. 2009. Case Studies on Measuring and Assessing Forest Degradation. A Local Level and Land Degradation Assessment Approach and a Case Study of Its Use in Senegal. Forest Resource Assessment Working Paper 174. Rome: FAO Forestry Department.

Burt, O. 1981. "Farm-Level Economics of Soil-Conservation in the Palouse Area of the Northwest." *American Journal of Agricultural Economics* 63 (1): 83–92.

CACILM. 2006. Republic of Uzbekistan National Programming Framework. Prepared by Republic of Uzbekistan UNCCD National Working Group. Draft, 28 February 2006. http://www.adb.org/Projects/CACILM/documents.asp.

Capoor, K., and F. Ambrosi. 2008. *State and Trends of the Carbon Market 2008: A Focus on Africa.* Washington, DC: World Bank.

Castro Filho, C., T. A. Cochrane, L. D. Norton, J. H. Caviglione, and L. P. Johansson. (2001). "Land Degradation Assessment: Tools and Techniques for Measuring Sediment Load." Third International Conference on Land Degradation and Meeting of the International Union of Soil Sciences,

Subcommission C: Soil and Water Conservation, Rio de Janeiro, September 17–21.

CBD (Convention on Biological Diversity). 2001. *The Value of Forest Ecosystems*. CBD Technical Series 4. Montreal.

Chomitz, K. M. 1996. "Roads, Land Use, and Deforestation: A Spatial Model Applied to Belize." *The World Bank Economic Review*. http://wber.oxfordjournals.org/content/10/3/487.abstract.

CIESIN (Center for International Earth Science Information Network). 2010. *Global Distribution of Poverty*. Palisades, NY. http://sedac.ciesin.columbia.edu/povmap/ds_info.jsp.

CIFOR (Center for International Forestry Research). 2008. *CIFOR's Strategy, 2008–2018: Making a Difference for Forests and People*. Bogor, Indonesia. www.cifor.cgiar.org/publications/pdf_files/Books/CIFORStrategy0801.pdf.

Clark, E.H., H.J. Haverkamp and W. Chapman. 1985. "Eroding Soils: The Off-Farm Impacts." Washington DC: The Conservation Foundation.

Clark, E. H. 1985. The Off-Site Costs of Soil Erosion. *Journal of Soil and Water Conservation* 40 (1): 19–22.

Clark, R. 1996. *Methodologies for the Economic Analysis of Soil Erosion and Conservation.* Center for Social and Economic Research on the Global Environment Working Paper, Global Export Control No. 96-13. Norwich, UK: University of East Anglia, Overseas Development Group.

Clarke, H.R. 1992. "The Supply of Non-Degraded Agricultural Land". *Australian Journal of Agricultural Economics* 36(1): 31-56.

Clay, D. C., F. U. Byiringiro, J. Kangasniemi, T. Reardon, B. Sibomana, L. Uwamariya, D. Tardif-Douglin. 1996. Promoting Food Security in Rwanda through Sustainable Agricultural Productivity: Meeting the Challenges of Population Pressure, Land Degradation, and Poverty. Food Security International Development Policy Syntheses 11425. East Lansing: Michigan State University, Department of Agricultural, Food, and Resource Economics.

Cleaver, K. M., and G. A. Schreiber. 1994. *Reversing the Spiral: The Population, Agriculture, and Environment Nexus in Sub-Saharan Africa.* Washington, DC: World Bank.

Coase, R. 1992. "The Institutional Structure of Production". *American Economic Review* 82(4): 713-719.

Cohen, J. A. 2005. *Intangible Assets: Valuation and Economic Benefit.* Hoboken, NJ: John Wiley and Sons.

Cohen, M. J., M. T. Brown, and K. D. Shephard. 2006. "Estimating the Environmental Costs of Soil Erosion at Multiple Scales in Kenya Using Emergy Synthesis." *Agriculture, Ecosystems, and Environment* 114: 249–269.

Collado, A., E. Chuvieco, and A. Camarasa. 2000. "Satellite Remote Sensing Analysis to Monitor Desertification Processes in the Crop-Rangeland Boundary of Argentina." *Journal of Arid Environments* 52: 121–133.

Colombo, S., N. Hanley and J. Calatrava-Requena. 2005. "Designing Policy for Reducing the Off-farm Effects of Soil Erosion Using Choice Experiments." *Journal of Agricultural Economics* 56(1): 81-95.

Colombo, S., J. Calatrava-Requena, and N. Hanley. 2003. "The Economic Benefits of Soil Erosion Control: An Application of the Contingent Valuation Method in the Alto Genil Basin of Southern Spain." *Journal of Soil and Water Conservation* 58 (6): 367–371.

Colson, E. 1974. Tradition and Contract: The Problem of Order. Chicago: Adeline Publishing.

Convery, F. and K. Tutu. 1990. "Evaluating the Costs of Environmental Degradation in Ghana. Applications of Economics in the Environmental Action Planning Process in Africa." University College Dublin, Dublin.

Cooke, P.; Köhlin, G. and W.F: Hyde (2008): Fuelwood, Forests and Community Management – Evidence from Household Studies. In: Environment and Development Economics 13: 103-135.

Cooper P., K. P. C. Rao, P. Singh, J. Dimes, P. S. Traore, K. Rao, P. Dixit, and S. Twomlow. 2009. "Farming with Current and Future Climate Change Risk: Advancing a 'Hypothesis of Hope' for Rainfed Agriculture in Semiarid Tropics." *Journal of SAT Agricultural Research* 7: 1–19.

Corcoran, E., C. Nellemann, E. Baker, R. Bos, D. Osborn, H. Savelli (eds). 2010. *Sick Water? The central role of wastewater management in sustainable development.* A Rapid Response Assessment. United Nations Environment Programme, UN-HABITAT, GRID-Arendal. www.grida.no.

Cornish G., B. Bosworth, C., Perry, and J. Burke. 2004. *Water Charging in Irrigated Agriculture. An Analysis of International Experience.* FAO Water Report 28. Rome: Food and Agriculture Organization of the United Nations.

Costanza, R., R. d'Arge, R. de Groot, S. Farberparallel, M. Grasso, B. Hannon, K. Limburg, et al. 1997. "The Value of the World's Ecosystem Services and Natural Capital." *Nature* 387: 253–260.

Coxhead, I. 1996. "Economic Modeling of Land Degradation in Developing Countries." Agricultural and Applied Economics Staff Paper Series. University of Wisconsin Madison.

Coxhead, I., and R. Oygard. 2007. *Land Degradation.* Draft submitted for Copenhagen Consensus 2008.

Craswell, E.T., U. Grote, J. Henao, and P. L. G. Vlek. 2004. "Nutrient Flows in Agricultural Production and International Trade: Ecological and Policy Issues." ZEF Discussion Paper on Development Policy 78. Center for Development Research, Bonn.

Croitoru, L., and M. Sarraf. 2010. The Cost of Environmental Degradation. Case Studies from the Middle East and North Africa. Washington DC: The World Bank.

Crosson, P. 1998. "The On-Farm Economic Costs of Soil Erosion." In *Methods for Assessment of Soil Degradation,* edited by R. Lal, W. H. Blum, C. Valentine, and B. A. Stewart, 495–511. Boca Raton, FL: CRC Press.

Cruz, W., H. Francisco, and Z. Tapawan-Conway. 1988. *The On-Site nd Downstream Costs f Soil Erosion.* Philippine Institute for Development Studies Working Paper No. 88-11. Los Baños: University of the Philippines.

Darghouth, S., C. Ward, G. Gambarelli, E. Styger, and J. Roux. 2008. *Watershed Management Approaches, Policies, and Operations: Lessons for Scaling Up.* Water Sector Bord Discussion Paper 1. Washington, DC: World Bank.

Dasgupta S., C. Meisner, and D. Wheeler. 2007. "Is Environmentally Friendly Agriculture Less Profitable for Farmers? Evidence on Integrated Pest Management in Bangladesh." *Applied Economic Perspectives and Policy* 29 (1): 103–118.

Davies, J. 2007. Total Economic Valuation of Kenyan Pastoralism. World Initiative for Sustainable Pastoralism. International Union for Conservation of Nature (IUCN), Nairobi.

Day, J. C., D. W. Hughes, and W. R. Butcher. 1992. "Soil, Water, and Crop Management Alternatives in Rainfed Agriculture in the Sahel: An Economic Analysis." *Agricultural Economics* 7: 267–287.

Defoer, T., A. Budelman, C. Toulmin, and S. Carter. 2000. "Building Common Knowledge: Participatory Learning and Action Research." In *Managing*

Soil Fertility in the Tropics. A Resource Guide for Participatory Learning and Action Research, edited by T. Defoer and A. Budelman. Amsterdam: Royal Tropical Institute.

De Fraiture, C., M. Giordano, and Y. Liao. 2008. "Biofuels and Implications for Agricultural Water Use: Blue Impacts of Green Energy." *Water Policy* 10 (1): 67–81.

De Graaf, J. 1996. "The Price of Soil Erosion: An Economic Evaluation of Soil Conservation and Watershed Development." PhD Thesis. Wageningen, The Netherlands: Landbouwuniversiteit.

De Graaff, J. 1993. Soil Conservation and Sustainable Land Use: An Economic Approach. Amsterdam: Royal Tropical Institute.

De Groot, R. S. 1992. Functions of Nature: Evaluation of Nature in Environmental Planning, Management, and Decision Making. Groningen, The Netherlands: Wolters-Noordhoff.

De Groot, R. S., M. A. Wilson, and R. M. Bouman. 2002. "A Typology for the Classification, Description, and Valuation of Ecosystem Functions, Goods, and Services." *Ecological Economics* 41 (3): 393–408.

Deininger, K. 2003. *Land Policies for Growth and Poverty Reduction.* World Bank Policy Research Report. Washington, DC: World Bank; Oxford, UK: Oxford University Press.

Deininger, K., and J. Chamorro. 2004. "Investment and Equity Effects of Land Regularization: The Case of Nicaragua." *Agricultural Economics* 30 (2): 101–116.

De Jong, S., and G. Epema. 2001. "Imaging Spectrometry Surveying and Modeling Land Degradation." In *Imaging Spectrometry. Basic Principles and Prospective Applications,* edited by F. Van der Meer and S. De Jong, S65–86. Dordrecht, The Netherlands: Kluwer Academic Publisher.

Den Biggelaar, C., R. Lal, K. Wiebe, and V. Breneman. 2003. "The Global Impact of Soil Erosion on Productivity: I: Absolute and Relative Erosion-Induced Yield Losses." *Advances in Agronomy* 81: 1–48.

De Jager a., D. Onduru and C. Walaga. 2004. Facilitated Learning in Soil Fertility Management: Assessing Potential of Low-External Inpt Technolgoies in East African Farming Systems. *Agricultural Systems* 79:205-223.

De Janvry, A.; Fafchamps, M. and E. Sadoulet 1991: Peasant Household behavior with missing markets: some paradoxes explained. *The Economic Journal* 101: 1400-1417.

Derpsch, R. and Friedrich, T. (2009) Global overview of Conservation Agriculture adoption. Invited Paper, 4th World Congress on Conservation Agriculture: Innovations for Improving Efficiency, Equity and Environment, 4–7 February. New Delhi: ICAR. On WWW at http://www.fao.org/ag/ca

De Pinto, A., M. Maghalaes, and C. Ringler. 2010. *Potential of Carbon Markets for Small Farmers*. IFPRI Discussion Paper 01004. Washington, DC: International Food Policy Research Institute.

Deshingkar, P., C. Johnson, and J. Farrington. 2005. "State Transfers to the Poor and Back: The Case of the Food-for-Work Program in India." *World Development* 33 (4): 575–591.

Devas, N., and U. Grant. 2003. "Local Government Decisionmaking: Citizen Participation and Local Accountability: Some Evidence from Kenya and Uganda." *Public Administration and Development* 23 (4): 307–316.

Diao, X., and D. B. Sarpong. 2007. *Cost Implications of Agricultural Land Degradation in Ghana***Fehler! Hyperlink-Referenz ungültig.**. IFPRI Discussion Paper 698. Washington, DC: International Food Policy Research Institute.

Diouf, E., and E. Lambin. 2001. "Monitoring Land-Cover Changes in Semiarid Regions: Remote Sensing Data and Field Observations in Ferlo, Senegal." *Journal of Arid Environments* 48: 129–148.

Dobos, E., J. Daroussin, and L. Montanrella. 2005. *An SRTM-Based Procedure to Delineate SOTER Terrain Units on 1:1 and 1:5 Million Scales.* Luxembourg: Office for Official Publications of the European Communities.

Doraiswamy, P. C., G. W. McCarty, E. R. Hunt Jr., R. S. Yost, M. Doumbia, and A. J. Franzluebbers. 2007. "Modeling Soil Carbon Sequestration n Agricultural Lands of Mali." *Agricultural Systems* 94: 63–74

Downing, T.E. 1991. "Vulnerability to Hunger in Africa: A Climate Change Perspective." *Global Environmental Change* 1(5): 365-380.

Drake, L., P. Bergstrom, and H. Svedsater. 1999. "Farmers' Attitude and Uptake." In *Countryside Stewardship: Farmers, Policies and Markets,* edited by G. Huylenbroeck and M. Whitby, 89–111. Pergamon Elsevier, Oxford, UK.

Dregne, H. E. 1977. Generalized Map of the Status of Desertification of Arid Lands, *Prepared by FAO of the United Nations, UNESCO and WMO for the 1977 United Nations Conference on Desertification* .

———. 1983. *Desertification of Arid Lands.* London: Harwood Academic Publishers.

Dregne, H., and N.-T. Chou. 1992. "Global desertification dimensions and costs. In *Degradation and restoration of arid lands.* edited by H. Dregne, 249–282. Lubbock: International Center for Arid and Semiarid Studies, Texas Tech University.

Duflo, E., and R. Pande. 2007. "Dams." *The Quarterly Journal of Economics* 122 (2): 601–646.

Duraiappah, A. 1998. "Poverty and Environmental Degradation: A Review and Analysis of the Nexus." *World Development* 26 (12): 2169–2179.

Eaton, D. 1996. "The Economics of Soil Erosion. A Model of Farm-Decision Making." Environmental Economics Program. Discussion paper 96-01. IID. FAO 1986. "Highlands Reclamation Study Ethiopia Final Report." Vol. I & II. Rome, Italy.

Edwards, C. 2010. *Ten Reasons to Cut Farm Subsidies.* Washington, DC: CATO Institute.

Ehui, S.K., T.W. Hertel, and P.V. Preckel. 1990. Forest Resource Depletion, Soil Dynamics, and agricultural productivity in the Tropics. *Journal of Environmental Economics and Management* 18: 136-154.

Ellis, E. C., and N. Ramankutty. 2008. "Putting People in the Map: Anthropogenic Biomes of the World." *Frontiers in Ecology and the Environment* 6: 439–447.

EI-Swaify S.A., E. W. Dangler and C. L. Armstrong. 1982. Soil Erosion by Water in the Tropics Reseaarch Extension Series 024, College Of Tropical Agriculture And Human Resources • University of Hawaii.

Enters, T. 1998. Methods for the Economic Assessment of the On- and Off-Site Impact of Soil Erosion. Bangkok: International Board for Soil Research and Management.

Eswaran, H., R. Lal, and P. Reich. 2001. "Land Degradation: An Overview." In Responses to Land Degradation. Proc. 2nd International Conference on Land Degradation and Desertification. Khon Kaen, Thailand, edited by E. Bridges, I. Hannam, L. Oldeman, F. Penning de Vries, S. Scherr, and S. Sompatpanit. New Delhi: Oxford Press.

Eswaran, H., P. Reich, and F. Beinroth. 1998. *Global Desertification Tension Zones.* In: D.E. Stott, R.H. Mohtar and G.C. Steinhardt (eds.): Sustaining the Global Farm. Selected Papers from the 10th International Soil Conservation Organization Meeting held May 24-29, 1999 at Purdue University and USDA-ARS National Soil Erosion Research Laboratory. http://www.tucson.ars.ag.gov/isco/isco10/SustainingTheGlobalFarm/P227-Eswaran.pdf.

Evans, J., and R. Geerken. 2004. "Discrimination between Climate and Human-Induced Dryland Degradation." *Journal of Arid Environments* 57: 535–554.

FAOSTAT. FAO statistics. Online at http://faostat.fao.org/.

FAO (United Nations Food and Agriculture Organization). n.d. *Cattle Ranching and Deforestation.* Livestock Policy Brief 3. Rome. ftp://ftp.fao.org/docrep/fao/010/a0262e/a0262e00.pdf.

———. 1979. A Provisional Methodology for Soil Degradation Assessment. Rome.

———. 1994. Land Degradation in South Asia: Its Severity, Causes and Effects upon the People. Rome.

———. 2001. Soil Fertility Management in Support of Food Security in Sub-Saharan Africa. Rome.

———. 2003. "Subsidies, Food Imports, and Tariffs. Key Issues for Developing Countries." Rome: FAO. www.fao.org/english/newsroom/focus/2003/wto2.htm.

———. 2008. Land Degradation Assessment in Drylands (LADA): Assessing the Status, Causes, and Impact of Land Degradation. Rome.

———. 2011. State of the World's Forests. Rome.

FAOSTAT Data. Online agricultural data. http://faostat.fao.org/.

Feather, P., D. Hellerstein, and L. Hansen. 1999. *Economic Valuation of Environmental Benefits and the Targeting of Conservation Programs: The Case of the CRP.* Resource Economics Division, Economic Research Service, Agricultural Economic Report No. 778. Washington, DC: U.S. Department of Agriculture.

Feder, G. 1987. "Land Ownership Security and Farm Productivity. Evidence from Thailand." *Journal of Development Studies* 24 (1): 16–30.

Filho, C., C. Cochrane, T. Norton, J. Caviglione, and L. Johanssion. 2001. "Land Degradation Assessment: Tools and Techniques for Measuring Sediment Load." Third International Conference on Land Degradation and Meeting of the International Union of Soil Sciences, Subcommission C: Soil and Water Conservation, Rio de Janeiro, September 17–21

Fischer, G., H. van Velthuizen, and F. Nachtergaele. 2000. *Global Agro-Ecological Zones Assessment: Methodology and Results.* Laxenburg, Austria: International Institute for Applied Systems Analysis.

Fisher, B., P. K. Turner, and P. Morling. 2009. "Defining and Classifying Ecosystem Services for Decision Making." *Ecological Economics* 68: 643–653.

Flaherty K., F. Murithi, W. Mulinge, and E. Njuguna. 2008. Kenya. Recent Developments in Public Agricultural Research. Country Note, June 2008, Agriculture Science and Technology Indicators (ASTI), International Food Policy Research Insitute (IFPRI) and Kenya Agricultural Research Institute (KARI).

Foster, R. 2006. "Methods for Assessing Land Degradation in Botswana." *Earth and Environment* 1: 238–276.

Foster, V., and C. Briceño-Garmendia. 2010. "Africa's Infrastructure: A Time for Transformation." Agence Française de Développement and the World Bank.

Freeman, A. M. 2003. *The Measurement of Environmental and Resource Values.* Washington, DC: Resources for the Future.

Gachimbi, L., A. de Jager, H. van Keulen, E. Thuranira, and S. Nandwa. 2002. *Participatory Diagnosis of Soil Nutrient Depletion in Semiarid Areas of Kenya.* London: International Institute for Environment and Development.

Gao, J., and Y. Liu. 2010. "Determination of Land Degradation Causes in Tongyu County, Northeast China via Land Cover Change Detection." *International Journal of Applied Earth Observation and Geoinformation* 12: 9–16.

Gao, J., Y. Zha, and S. Ni. 2001. "Assessment of the Effectiveness of Desertification Rehabilitation Measures in Yulin, Northwestern China, Using Remote Sensing." *International Journal of Remote Sensing* 22 (18): 3783–3795.

Gavian, S., and M. Fafchamps. 1996. "Land Tenure and Allocative Efficiency in Niger." *American Journal of Agricultural Economics* 78: 460–471.

Gebremedhin, B., S. M. Swinton, and Y. Tilahun. 1999. "Effects of Stone Terraces on Crop Yields and Farm Profitability: Results of On-Farm Research in Tigray, Northern Ethiopia." *Journal of Soil and Water Conservation* 54 (3): 568–573.

Geist, H. 2005. The Causes and Progression of Desertification. Great Britain: Ashgate .

Geist, H. J., and E. F. Lambin. 2004. "Dynamical Causal Patterns of Desertification." *BioScience* 54 (9): 817–829.

Gibson, C., J. Williams, and E. Ostrom. 2005. "Local Enforcement and Better Forests." *World Development* 33 (2): 273–284.

Giller, K., G. Cadisch, C. Ehaliotis, and E. Adams. 1997. "Building Soil Nitrogen Capital in Africa." In *Replenishing Soil Fertility in Africa,* edited by R. Buresh, P. Sanchez, and F. Calhoun. Madison, WI: Soil Science Society of America.

Glantz, M. H., and N. Orlovsky. 1983. "Desertification: A Review of the Concept Desertification." *Desertification Control Bulletin* 9: 15–22. Accessed online at http://www.ciesin.org/docs/002-479/002-479.html.

Görlach, B., R. Landgrebe-Trinkunaite, and E. Interwies. 2004. *Assessing the Economic Impacts of Soil Degradation. Volume I: Literature Review.* Study commissioned by the European Commission, DG Environment (Study Contract ENV.B.1/ETU/2003/0024). Berlin: Ecologic.

Gregersen, H. M., K. N. Brooks, J. A. Dixon, and L. S. Hamilton. 1987. *Guidelines for Economic Appraisal of Watershed Management Practices* (FAO Conservation Guide 16). Rome: FAO.

Grepperud, S. 1996. "Population Pressure and Land Degradation: The Case of Ethiopia." *Journal of Environmental Economics and Management* 30 (1): 18–33.

Grepperud, S. 1991. "Soil conservation as an investment in land." Discussion Paper 163. Oslo, Statistiques Norvège.

Gretton, P., and U. Salma. 1997. "Land Degradation: Links to Agricultural Output and Profitability." *The Australian Journal of Agricultural and Resource Economics* 41 (2): 209–225.

Grey, D., and C. Sadoff. 2006. "Water for Growth and Development: A Framework for Analysis." Theme Document of the Fourth World Water Forum, Mexico City, March.

Grohs, F. 1994. *Economics of Soil Degradation, Erosion and Conservation: A Case Study of Zimbabwe.* In: Kiel: Arbeiten zur Agrarwirtschaft in Entwicklungsländern. Wissenschaftsverlag Vauk Kiel KG, Germany.

Kurzman, C., R. Werun, and R. E. Burkhart. 2002. "Democracy's effect on economic growth: a pooled time-series analysis, 1951–1980." *Studies in Comparative International Development.* 37(1):3–33.

Gunatilake, H. M., and G. R. Vieth. 2000. "Estimation of On-Site Cost of Soil Erosion: A Comparison of Replacement and Productivity Change Method." *Journal of Soil and Water Conservation* 55 (2): 197–204.

Hajkowicz, S., and M. D. Young. 2002. "An Economic Analysis of Revegetation for Dryland Salinity Control on the Lower Eyre Peninsula in South Australia." *Land Degradation and Development* 13 (5): 417–428.

Hansen, L. T., V. E. Breneman, C. W. Davison, and C. W. Dicken. 2002. "The Cost of Soil Erosion to Downstream Navigation." *Journal of Soil and Water Conservation* 57 (2): 205–212.

Hansen, L., and D. Hellerstein. 2007. "The Value of the Reservoir Services Gained with Soil Conservation." *Land Economics* 83 (3): 285–301.

Häusler N. 2009. The challenges of a success story Posada Amazonas, Peru: A joint venture between a Tour Operator and a local Community. Online: http://www.mascontour.net/Media/Veroeffentlichungen/Posada%20Amazonas.pdf.

Hayami, Y. and V. W. Ruttan. 1970. Factor Prices and Technical Change in Agricultural Development: The United States and Japan, 1880-1960. *The Journal of Political Economy* 78, 1115-1141.

Hazell, P. 2010. "An Assessment of the Impact of Agricultural Research in South Asia since the Green Revolution." In *Handbook of Agricultural Economics.* Vol. 4, edited by P. Pingali and R. Evenson, 3469–3530.

Heffer P., and M. Prud'homme. 2009. "Fertilizer Outlook 2009–2013." 77th International Fertilizer Industry Association Annual Conference, Shanghai, May 25–27.

Hein, L. 2002. *Checklist of Potential Costs of Land Degradation and the Benefits of Mitigation Measures.* Wageningen, Masstricht: International Center for Environmental Assessments, Foundation for Sustainable Development, International Centre for Integrative Studies.

———. 2007. "Assessing the Costs of Land Degradation: A Case Study for the Puentes Catchment, Southeast Spain." *Land Degradation and Development* 18: 631–642.

Hein, L., and N. de Ridder. 2006. "Desertification in the Sahel: A Reinterpretation." *Global Change Biology* 12: 751–758.

Helldén, U., and C. Tottrup. 2008. "Regional Desertification: A Global Synthesis." *Global and Planetary Change* 64: 169–176.

Henao, J., and C. Baanante. 1999. *Estimating Rates of Nutrient Depletion in Soils of Agricultural Lands of Africa.* Muscle Shoals, AL: International Fertilizer Development Center.

Henning J., G. Lacefield, M. Rasnake, R. Burris, J. Johns, K. Johnson, and L. Turner. 2000. *Rotational Grazing,* University of Kentucky Cooperative

Extension ID-143. Lexington: University of Kentucky. http://search.uky.edu/Ag/AnimalSciences/pubs/id143.pdf

Herrmann, S., A. Anyamba, and C. J. Tucker. 2005. "Recent Trends in Vegetation Dynamics in the African Sahel and Their Relationship to Climate." *Global Environmental Change* 15 (4): 394–404.

Herrmann, S., and C. Hutchinson. 2005. "The Changing Contexts of the Desertification Debate." *Journal of Arid Environments* 63: 538–555.

Hill, J., J. Mégier, and W. Mehl. 1995. "Land Degradation, Soil Erosion, and Desertification Monitoring in Mediterranean Ecosystems." *Remote Sensing Review* 12: 107–130.

Hill, J., M. Stellmes, T. Udelhoven, A. Röder, and S. Sommer. 2008. "Mediterranean Desertification and Land Degradation. Mapping Related Land Use Change Syndromes Based on Satellite Observations." *Global and Planetary Change* 64: 146–157.

Hill, J., G. Tsioulis, P. Kasapidis, T. Udelhoven, and C. Diemer. 1998. "Monitoring 20 Years of Increased Grazing Impact on the Greek Island of Crete with Earth Observation Satellites." *Journal of Arid Environments* 39: 165–178.

Hobbs P. 2007. Conservation Agriculture: What Is It and Why Is It Important for Future Sustainable Food Production? *Journal of Agricultural Sciences.* 145: 127–137.

Hoffman G. 1986. Irrigation Management for Soil Salinity Control: Theories and Tests. *Soil Science Society of America Journal* 50:1552-1560

Holden, S. and H. Lofgren. 2005. Assessing the Impacts of Natural Resource Management Policy Interventions with a Village General Equilibrium Model. In *Natural Resource Management in Agriculture: Methods for Assessing Economic and Environmental Impacts,* edited by B. Shiferaw, H. Ade Freeman and S. Swinton. CABI Publishing, pp. 295-318.

Holden, S., and B. Shiferaw. 2004. "Land Degradation, Drought, and Food Security in a Less-Favored Area in the Ethiopian Highlands: A Bio-Economic Model with Market Imperfections." *Agricultural Economics* 30: 31–49.

Holden, S., B. Shiferaw, and J. Pender. 2004. "Nonfarm Income, Household Welfare, and Sustainable Land Management in a Less-Favored Area in the Ethiopian Highlands." *Food Policy* 29: 369–392.

Holmes, T. P. 1988. "The Offsite Impact of Soil Erosion on the Water Treatment Industry." *Land Economics* 64 (4): 356–366.

Hopkins, J., D. Southgate, and C. Gonzalez-Vega. 1999. "Rural Poverty and Land Degradation in El Salvador." Paper Presented at the Agricultural and Applied Economics Associations Annual Meeting. Abstract in *American Journal Agricultural Economics* 81 (December 1999).

Huang, J., and S. Rozelle. 1995. "Environmental Stress and Grain Yields in China." *American Journal of Agricultural Economics* 77: 853-864.

Huete, A. (1988). "A Soil-Adjusted Vegetation Index (SAVI)." *Remote Sensing of Environment* 25: 295–309.

Huete, A., C. Justice and H. Liu. 1994. "Development of Vegetation and Soil Indices for MODIS-EOS.) *Remote Sensing Environment* 49: 224-234.

ICRAF (International Center for Research in Agroforestry). 2008. *Agroforestry for Food Security and Healthy Ecosystems.* World Agroforestry Center Annual Report. Nairobi: World Agroforestry Center.

IMF (International Monetary Fund). 2010. World Economic Outlook Data. Online at: http://www.imf.org/external/pubs/ft/weo/2011/01/weodata/

Ishaq M., A. Hassan, M. Saeed, M. Ibrahim and R. Lal. 2001.Subsoil compaction effects on crops in Punjab, Pakistan: I. Soil physical properties and crop yield. *Soil and Tillage Research* 59(1-2): 57-65.

Jayasuriya, R. T. 2003. "Measurement of the Scarcity of Soil in Agriculture." *Resources Policy* 29: 119–129.

Jensen, J. 2000. Remote Sensing of the Environment. An Earth Resource Perspective. New Jersey: Prentice-Hall.

Johnson, D., and L. Lewis. 2007. *Land Degradation: Creation and Destruction.* 2nd ed.New York, USA: Rowman and Littlefield.

Jones, J.W., G. Hoogenboom, C.H. Porter, K.J. Boote, W.D. Batchelor, L.A. Hunt, P.W. Wilkens, U. Singh, A.J. Gijsman, and J.T. Ritchie. 2003. The DSSAT cropping system model. *European Journal of Agronomy* 18:235-265.

Junge B., G. Dercon; R. Abaidoo; D. Chikoye and K. Stahr. 2008. Estimation of medium-term soil redistribution rates in Ibadan, Nigeria, by using the 137Cs technique. Tropentag 2008 University of Hohenheim, October 7-9, 2008 Conference on International Research on Food Security, Natural Resource Management and Rural Development.

Justice, C., E. Vermote, J. Townshend, R. Defries, D. Roy, D. K. Hall, V. V. Salomonson, et al. 1998. "The Moderate Resolution Imaging Spectroradiometer (MODIS): Land Remote Sensing for Global Change

Research." *IEEE Transactions on Geoscience and Remote Sensing* 36 (4): 1128–1249.

Kabubo-Mariara, J. 2007. "Land Conservation and Tenure Security in Kenya: Boserup's Hypothesis Revisited." *Ecological Economics* 64: 25–35.

Kaliba, A. R., and T. Rabele. 2004. "Impact of Adopting Soil Conservation Practices on Wheat Yield in Lesotho." In: Bationo (ed), A. *Managing Nutrient Cycles to Sustain Soil Fertility in Sub-Saharan Africa,* Tropical Soil Biology and Fertility Institute of CIAT.

Kassam, A., T. Friedrich, F. Shaxson, and J. Pretty. 2009. "The Spread of Conservation Agriculture: Justification, Sustainability, and Uptake." *International Journal of Agricultural Sustainability* 7(4): 292–320.

Kassie, M., J. Pender, M. Yesuf, G. Kohlin, R. Bluffstone, and E. Mulugeta. 2008. "Estimating Returns to Soil Conservation Adoption in the orthern Ethiopian highlands." *Agricultural Economics* 38: 213–232.

Katyal, J. C., and P. L. Vlek. 2000. *Desertification: Concept, Causes, and Amelioration.* Bonn: Center for Development Research.

Kerr, J. 2007. Watershed Management: Lessons from Common Property Theory. *International Journal of the Commons* 1(1): 89-110.

Kherallah M., C. Delgado, E. Gabre-Madhin, N. Minot, and M. Johnson. 2002. *The Road Half Traveled: Agricultural Market Reform in Sub-Saharan Africa.* IFPRI Food Research Report. Washington, DC: International Food Policy Research Institute.

Khusamov, R., Kienzler, K., A. Saparov, M. Bekenov, B. Kholov, M. Nepesov, R. Ikramov, A. Mirzabaev, and R. Gupta 2009. Sustainable Land Management Research Project 2007- 2009.Final Report – Part III (Socio-Economic Analysis). ICARDA Central Asia and Caucasus Program. Tashkent, Uzbekistan. 238 pp.

King, D. A., J. A. Sinden. 1988. "The influence of Soil Conservation on farm land values."*Land Economics* 64(3): 242-255.

Kiss, A. 2004. "Is Community-Based Ecotourism a Good Use of Biodiversity Conservation Funds?" *Trends in Ecology & Evolution* 19 (5): 232–237.

Kiunsi, R., and M. Meadows. 2006. "Assessing Land Degradation in the Monduli District, Northern Tanzania." *Land Degradation and Development* 17: 509–525.

Klintenberg, P., and M. Seely. 2004. "Land Degradation Monitoring in Namibia. A First Approximation." *Environmental Monitoring and Assessment* 99: 5–21.

Koning, N. and E. Smaling. 2005. "Environmental Crisis or 'Lie of the Land'? The debate on soil degradation in Africa." *Land Use Policy*, 22(1): 1-9.

Krausse, M., C. Eastwood, and R. R. Alexander 2001. Muddied Waters: Estimating the National Economic Cost of Soil Erosion and Sedimentation in New Zealand. Lincoln Landcare Research. Palmerston North, New Zealand.

Kruseman, G., H. Hengsdijk, R. Ruben, P. Roebeling, and J. Bade. 1997. *Farm Household Modeling System: Theoretical and Mathematical Description.* DLV Report No. 7. Wageningen, The Netherlands.

Kumar, P., and S. Mittal. 2006. "Agricultural Productivity Trends in India: Sustainability Issues." *Agricultural Economics Research Review* 19: 71–88.

Kumar, R., R. D. Singh, and K. D. Sharma. 2005. "Water Resources of India." *Current Science* 89 (5): 794–811.

Kuppannan, P., and S. K. Devarajulu. 2009. "Impacts of Watershed Development Programmes: Experiences and Evidences from Tamil Nadu." Munich Research Papers in Personal Economics Personal Archive. http://mpra.ub.uni-muenchen.de/18653/.

LADA, 2008: Mapping Land Use Systems at global and regional scales for Land Degradation Assessment Analysis. In: Nachtergaele F. and M. Petri: LADA Technical report n.8, version 1.1.

Laflen, J., L. J. Lane, and G. R. Foster. 1991. "WEPP—A Next Generation of Erosion Prediction Technology." *Journal of Soil Water Conservation* 46 (1): 34–38.

Lal. R. 2004. "Soil Carbon Sequestration Impacts on Global Climate Change and Food Security." *Science* 304: 1623–1627.

——— 1987. "Effects of Erosion on Crop Productivity." *Critical Review in Plant Sciences* 5 (4): 303–367.

——— 1981. "Soil Erosion Problems on Alfisols in Western Nigeria. VI. Effects of Erosion on Experimental Plots." *Geoderma* 25:215.

Lambin, E. 2002. "The Interplay between International and Local Processes Affecting Desertification." In *Global Desertification: Do Humans Cause Deserts?* edited by J. Reynolds, and D. Stafford Smith, S387–400. Berlin: Dahlem University Press.

Lapar, M. L., and S. Pandey. 1999. "Adoption of Soil Conservation: The Case of Philippine Uplands." *Agricultural Economics* 21: 241–256.

Latham, J. 2011. “Designing a Global Monitoring Network for Agriculture, Ecosystem Services and Livelihoods.” Paper presented at the Bill and Melinda Gates Foundation Agricultural Monitoring Workshop, Seattle, WA, January 31–February 2.

Lebedys, A. 2008. Contribution of the Forestry Sector to National Economies, 1990–2006. Rome: FAO Forest Economics and Policy Division.

Leberl, F. 1990. *Radargrammatric Image Processing.* Norwood, MA: Artech House.

Leemans, R., and A. Kleidon. 2002. “Regional and Global Assessment of the Dimensions of Desertification.” In *Global Desertification: Do Humans Cause Deserts?* edited by J. Reynolds and D. Stafford Smith, 215–232. Berlin: Dahlem University Press.

Le Houérou, H. 1984. “Rain-Use Efficiency: A Unifying Concept in Arid-Land Ecology.” *Journal of Arid Environments* 7: 213–247.

Lepers, E., E. F. Lambin, A. C. Janetos, R. De Fries, F. Achard, N. Ramankutty, and R. J. Scholes. 2005. “A Synthesis of Information on Rapid Land-cover Change for the Period 1981–2000.” *BioScience* 55 (2): 115–124.

Lillesand, T., R. Kiefer, and J. Chipman. 2004. “Microwave and Lidar Sensing.” In *Remote Sensing and Image Interpretation,* edited by T. Lillesand, R. Kiefer, and J. Chipman, S638–738. New York: John Wiley & Sons.

Lopez, R. 1997, Land Titles and Farm Productivity in Honduras, World Bank, Washington DC, mimeo.

Louviere, J., D. Hensher, and J. Swait. 2000. Stated Choice Methods. Analysis and Application in Marketing, Transportation and Environmental Valuation. Cambridge University Press.

Lu, D., M. Batistella, P. Mausel, and E. Moran. 2007. “Mapping and Monitoring Land Degradation Risks in the Western Brazilian Amazon Using Multitemporal Landsat TM/ETM+ Images.” *Land Degradation and Development* 18: 41–54.

Lutz, E., S. Pagiola, and C. Reiche. 1994. “The Costs and Benefits of Soil Conservation: The Farmers’ Viewpoint.” *The World Bank Research Observer* 9 (2): 273–295.

MA (Millenium Ecosystem Assessment). 2005a. “Dryland Systems.” In *Ecosystem and Well-Being: Current State and Trends,* edited by R. Hassan, R. Scholes, and N. Ash, 623–662. Washington, DC: Island Press.

———. 2005b. Living Beyond Our Means: Natural Assets and Human Well-Being. Washington, DC: Island Press.

MA 2010.Millenium Development Goals Report, 2010. United Nations:52-63

Ma, H., and H. Ju. 2007. "Status and Trends in Land Degradation in Asia." In *Climate and Land Degradation,* edited by M. V. Sivakumar and N. Ndiang´ui, 55–65. Berlin Heidelberg, Germany: Springer-Verlag.

Maas, E.V. and G.J. Hoffman. 1977. Crop Salt Tolerance - Current Assessment. *Journal of Irrigation and Drainage Engineering* 103(IR2):115-134

Magrath, W., and P. L. Arens. 1989. *The Cost of Soil Erosion on Java: A Natural Resource Accounting Approach.* Environment Department Working Paper 18. Washington, DC: World Bank.

Mainguet, M., and G. G. da Silva. 1998. "Desertification and Drylands Development: What Can Be Done?" *Land Degradation and Development* 9 (5): 375–382.

Mazzucato, V., and D. Niemeijer. 2001. *Overestimating Land Degradation, Underestimating Farmers in the Sahel.* London: International Institute for Environment and Development.

McConnell, K. 1983. "An Economic Model of Soil Conservation." *American Journal of Agricultural Economics* 65: 83–89.

McIntire, J. (1994). "A review of the soil conservation sector in Mexico. In: Economic and institutional analyses of soil conservation projects in Central America and the Caribbean." In *A CATIE World Bank Project,* edited by E. Lutz, S. Pagiola, and C. Reiche. World Bank Environment Paper 8. Washington, D.C.: The World Bank.

McKenzie, C. 1994. "Degradation of Arable Land Resources. Policy Options and Considerations within the Context of Rural Restructuring." Paper presented at the Land and Agriculture Policy Centre (LAPC) Workshop, 30–31 March 1994, Johannesburg, South Africa.

McKinsey and Company. 2009. Charting Our Future. Economic Frameworks to Inform Decision-Making. 2030 Water Resources Group. http://www.mckinsey.com/clientservice/water/charting_our_water_future.aspx

Meadows, M., and T. Hoffman. 2003. "Land Degradation and Climate Change in South Africa." *The Geographical Journal* 2: 168–177.

Meinzen-Dick, R. S., and E. Mwangi. 2008. "Cutting the Web of Interests: Pitfalls of Formalizing Property Rights." *Land Use Policy* 26 (1): 36–43.

Mekuria M., S. Waddington. 2002. Initiatives to encourage farmer adoption of soil fertility technolcoies for maize-based cropping systems in Southern Africa. In: Barrett,C., F. Place, and A. Abdillahi (eds). *Natural Resource*

Management in African Agriculture: Understanding and improving current practices, Wallingford: CAB International.

Mengistu, Y. A. 2011. "The Effect of Policy Incentives and Technology on Sustainable Land Management and Income of Small Farm Households: A Bioeconomic Model for Anjeni Area, North Western Ethiopia." Dissertation, University of Bonn, Cuvilier Verlag Göttingen, Germany.

Migot-Adholla, S., P. Hazell, B. Blarel, and F. Place. 1991. "Indigenous Land Right Systems in Sub-Saharan Africa: A Constraint on Productivity?" *The World Bank Economic Review* 5 (1): 155–175.

Miranda L. M. 1992. "Landowner Incorporation of On-site Soil Erosion Costs: an Application to the Conservation Reserve Program." *American Journal of Agricultural Economics* 74: 434-43.

Mitchell R., and R. T. Carson. 1989. "Using .Surveys To Value Public Goods: The Contingent valuation Method." Resources for the Future, Washington, D.C.

MNP (Netherlands Environmental Assessment Agency)/OECD (Organization of Economic Cooperation and Development). 2007. *Background Report to the "OECD Environmental Outlook to 2030." Overviews, Details, and Methodology of Model-Based Analysis.* Bilthoven, The Netherlands: Netherlands Environmental Assessment Agency; Paris: Organization of Economic Cooperation and Development.

Montanarella, L. 2007. "Trends in Land Degradation in Europe." In *Climate and Land Degradation,* edited by M. V. Sivakumar and N. Ndiang´ui, 83–105. Berlin Heidelberg, Germany: Springer-Verlag.

Moore, W. B., and Bruce A. McCarl. 1987. "Off-Site Costs of Soil Erosion: A Case Study in the Willamette Valley." *Western Journal of Agricultural Economics* 12: 42- 49.

Mountian Forum Bulletin. 2010. *Mountain Forum Bulletin* 19(1). Keenan (Ed.). Mountain Forum Secretariat, ICRAF, IUCN, ICIMOD and the World Bank (Publishers).

Muchena, F. N., D. D. Onduru, G. N. Gachini, and A. de Jager. 2005. "Turning the Tides of Soil Degradation in Africa: Capturing the Reality and Exploring Opportunities." *Land Use Policy* 22: 23–31.

Mullen, J. D. 2001. *An Economic Perspective on Land Degradation.* New South Wales Agriculture, Orange. Economic Research Report No. 9.

Mulvaney, R. L., S. A. Khan, and T. R. Ellsworth. 2009. "Synthetic Nitrogen Fertilizers Deplete Soil Nitrogen: A Global Dilemma for Sustainable Cereal Production." *Journal of Environmental Quality* 38: 22951.

Murphee, C. E., and K. C. McGregor. 1991. "Runoff and Sediment Yield from a Flatland Watershed in soybeans." *Transactions of ASABE* 34: 407–411.

Nachtergaele, F., R. Biancalani, S. Bunning, and H. George. 2010. *Land Degradation Assessment: The LADA Approach.* 19th World Congress of Soil Science, Brisbane, Australia.

Nachtergaele, F., and M. Petri. 2008. Mapping Land Use Systems at Global and Regional Scales for Land Degradation Assessment Analysis. Version 1.0. Technical report n.8 of the LADA FAO / UNEP Project.

Nachtergaele, F., M. Petri, R. Biancalani, G. Van Lynden, and H. Van Velthuizen. 2010. Global Land Degradation Information System (GLADIS). Beta Version. An Information Database for Land Degradation Assessment at Global Level. Land Degradation Assessment in Drylands Technical Report, no. 17. FAO, Rome, Italy.

Napier, T. L. 1991. "Factors Affecting Acceptance and Continued use of Soil Conservation Practices in Developing Societies: A Diffusion Perspective". *Agriculture, Ecosystems and Environment* 36: 127-140.

Nardi, D. 2005. "The Green Monks of Burma: The Role of Monks and Buddhist Civil Society in Environmental Conservation in Burma." *The Georgetown University Journal of the Environment* 7: 5–14.

Ndegwa, S., and B. Levy. 2004. "The Politics of Decentralization in Africa: A Comparative Analysis" In *Building State Capacity in Africa: New Approaches, Emerging Lessons,* edited by B. Levy and S. A. Kpundeh.

Ndiritu, C. 1992. A Search for Strategies for Sustainable Dry Land Cropping in Semiarid Eastern Kenya. Proceedings of a symposium held in Nairobi, Kenya, 10-11 December 1990, ACIAR Proceedings, no. 41, edited by M. Probert. Canberra, Australia.

Nelson, R. A., R.A. Cramb, and M.A. Mamicpic. 1998. "Erosion/Productivity Modeling of Maize Farming in the Philippine uplands: Part III: Economic Analysis of Alternative Farming Methods." *Agricultural Systems* 58(2): 165-183.

Nelson, A. 2008. "Estimated Travel Time to the Nearest City of 50,000 or More People in the Year 2000." Global Environmental Monitoring Unit. Ispra, Italy: Joint Research Centre of the European Commission.

Niemeijer, D. 1999. "Environmental Dynamics, Adaptation, and Experimentation in Indigenous Sudanese Water Harvesting." In *Biological and Cultural Diversity: The Role of Indigenous Agricultural Experimentation in Development: 64-79.*, edited by S. F. G. Prain, London, UK.

Nkonya, E., J. Pender, K. Kaizzi, E. Kato, S. Mugarura, H Ssali, and J. Muwonge. 2008a. Linkages between land management, land degradation, and poverty in Sub-Saharan Africa: The case of Uganda. IFPRI Research Report #159, Washington D.C.

Nkonya, E., P. Gicheru, J. Woelcke, B. Okoba, D. Kilambya, and L. N. Gachimbi. 2008b. *On-Site and Off-Site Long-Term Economic Impacts of Soil Fertility Management Practices.* IFPRI Discussion Paper 778. Washington, DC: International Food Policy Research Institute.

Nkonya E., J. Pender, and E. Kato. 2008. "Who Knows, Who Cares? The Determinants of Awareness, Enactment, and Compliance with Natural Resource Management Regulations in Uganda." *Environment and Development Economics* 13 (1): 79–109.

Nkonya E., D. Phillip, T. Mogues, J. Pender, and E. Kato. 2010. *From the Ground Up: Impacts of a Pro-poor Community-Driven Development Project in Nigeria.* IFPRI Research Monograph. Washington, DC: International Food Policy Research Institute.

Nkonya E., F. Place, J. Pender, Majaliwa Mwanjololo, A. Okhimamhe E. Kato; S. Crespo; J. Ndjeunga and S. Traore 2011. Climate Risk Management through Sustainable Land Management in Sub-Saharan Africa. IFPRI Discussion Paper (forthcoming).

Nori M., M. Taylor, and A. Sensi. 2008. Browsing on fences: pastoral land rights, livelihoods and adaptation to climate change. IIED Drylands Series #148 , London.

Norris, E., and S. Batie. 1987. "Virginia Farmers' Soil Conservation Decisions: An Application of Tobit Analysis." *Southern Journal of Agricultural Economics* 19 (1): 88–97.

North, D.C. 1990. Institutions, Institutional Change, and Economic Performance. Cambridge: Cambridge University.

North, D.C. 1991. "Institutions". *The Journal of Economic Perspectives* 5(1): 97-112.

Norse, D. and Saigal, R.. 1992. "National economic cost of soil erosion--the case of Zimbabwe." Paper prepared for the CIDIE workshop on Environmental Economics and Natural Resource Management in Developing Countries, World Bank, Washington, DC, 22-24 January 1992. FAO, Rome.

Nyangena, W., and G. Köhlin. 2008. *Estimating Returns to Soil and Water Conservation Investments: An Application to Crop Yield in Kenya.*

Enviroment for Development Discussion Paper Series 08-32. Resources for the Future. Göteborg. Sweden.

OECD (Organization for Economic Cooperation and Development). 2010. Creditor Reporting System Online Database on Aid Activities. www.oecd.org/dac/stats/idsonline.

Öesterreichische Bundesforste AG. 2009. Assessment of the Value of the Protected Area System of Ethiopia, "Making the Economic Case.". Prepared by the Österreichische Bundesforste AG for Ehtiopian Wildlife Authority Sustainable Development of the Protected Area System of Ethiopia (SDPASE) Project.

Oldeman, L. 1996. Global and Regional Databases for Development of State Land Quality Indicators: The SOTER and GLASOD Approach. ISRIC Report 96/2.Wageningen, The Netherlands.

———. 1998. *Soil Degradation: A Threat to Food Security?* ISRIC Report 98/01. Wageningen, The Netherlands: International Soil Reference and Information Center.

Oldeman, L. R., R. T. Hakkeling, and G. Sombroek. 1991a. *Status of Human-Induced Soil Degradation* 2nd ed. Wageningen, The Netherlands: International Soil Reference and Information Center.

———. 1991b. World map of the status of human-induced soil degradation. an explanatory note. Global Assessment of Soil Degradation (GLASOD). (I. S. (ISRIC), & U. N. (UNEP), Hrsg.) Wageningen.

Oldeman, R. 2002. "Assessment of Methodologies for Drylands Land Degradation Assessment." First Meeting of Technical Advisory Group and Steering Committee of the Land Degradation Assessment in Drylands. FAO, Rome, January 22–25. Wageningen, The Netherlands.

Ole-Lengisugi, M. N. 1998. "Indigenous Knowledge and Skills in Combating Desertification and Drought." www.worldbank.org/afr/ik/ikpacks/environment.htm#rutr.

Olson, M. Jr. 1996. "Big Bills Left on the Sidewalk: Why Some Nations Are Rich, and Others are Poor." *Journalof Economic Perspectives* 10(2): 3-24.

Omuto, C., and R. Vargas. 2009. "Combining Pedometrics, Remote Sensing and Field Observations for Assessing Soil Loss in Challenging Drylands: A Case Study of Northwestern Somalia." *Land Degradation and Development* 20: 101–115.

Ostrom, E. 1990. Governing the Commons: The Evolution of Institutions for Collective Action: Political Economy of Institutions and Decisions. New York: Cambridge University Press.

Ostrom, E., and H. Nagendra. 2006. "Insights on Linking Forests, Trees, and People from the Air, on the Ground, and in the Laboratory." *Proceedings of the National Academy of Sciences* 103 (51): 19224–19231.

Padwick, G. 1983. "Fifty Years of Experimental Agriculture. II. The Maintenance of Soil Fertility in Tropical Africa: A Review." *Experimental Agriculture* 19: 293–310.

Pagiola, S. 1996. "Price Policy and Returns to Soil Conservation in Semiarid Kenya." *Environmental and Resource Economics* 8 (3): 255–271.

———. 2008. "Payments for Environmental Services in Costa Rica." *Ecological Economics* 65 (4): 712–724.

Pagiola, S., K. von Ritter, and J. Bishop. 2004. *Assessing the Economic Value of Ecosystem Conservation.* The World Bank Environment Department, Paper 101.Washington, DC: World Bank.

Pandey, D. 2007. "Multifunctional Agroforestry Systems in India." *Current Science* 92 (4): 455–463

Partap, T., and H. R. Watson. 1994. *Sloping Agricultural Land Technology (SALT): A Regenerative Option for Sustainable Mountain Farming.* Kathmandu: International Center for Integrated Mountain Development.

Paudel, G. S., and G. B. Thapa. 2004. "Impact of Social, Institutional, and Ecological Factors on Land Management Practices in Mountain Watersheds of Nepal." *Applied Geography* 24 (1): 35–55.

Pender J., A. Mirzabaev and E. Kato. 2009. Economic Analysis of Sustainable Land Management Options in Central Asia. IFPRI mimeo.

Pender, J. L. 2009. "Food Crisis & Land: The World Food Crisis, Land Degradation, and Sustainable Land Management: Linkages, Opportunities, and Constraints." German Organization for Technical Cooperation and TerrAfrica. www2.gtz.de/dokumente/bib/gtz2009-0196en-food-crisis-land.pdf.

———. 1992. "Credit Rationing and Farmers' Irrigation Investments in Rural South India: Theory and Evidence." Ph.D. Diss., Food Research Institute, Stanford University. Processed.

Pender, J., E. Nkonya, P. Jagger, D. Sserunkuuma, and H. Ssali. 2006. "Strategies to Increase Agricultural Productivity and Reduce Land Degradation in Uganda: An Econometric Analysis." In *Strategies for Sustainable Land Management in the East African Highlands*, edited by J. P. Pender and S. Ehui, S165–190. Washington, DC: International Food Policy Research Institute.

Pender, J. L., and B. Gebremedhin. 2006. "Land Management, Crop Production, and Household Income in the Highlands of Tigray, Northern Ethiopia: An Econometric Analysis." In *Strategies for Sustainable Land Management in the East African Highlands,* edited by J. P. Pender and S. Ehui, S107–138. Washington, DC: International Food Policy Research Institute.

Pender, J. L., and J. M. Kerr. 1998. "Determinants of Farmers' Indigenous Soil and Water Conservation Investments in Semiarid India." *Agricultural Economics* 19: 113–125.

Pender, J.; Place, F. and S. Ehui. 2006. Strategies for Sustainable Land Management in the East African Highlands. Washington, D.C. International Food Policy Research Institute.

Pender, J. L., C. Ringler, M. Magalhaes, and F. Place. 2009. "The Role of Sustainable Land Management for Climate Change Adaptation and Mitigation in Sub-Saharan Africa." TerrAfrica. http://startinternational.org/library/items/show/5.

Pimentel, D. 2006. "Soil Erosion: A Food and Environmental Threat." *Environment, Development, and Sustainability* 8: 119–137.

Pimentel, D., C. Wilson, C. McCullum, R. Huang, P. Dwen, J. Flack, Q. Tran, T. Saltman, and B. Cliff. 1997. "Economic and Environmental Benefits of Biodiversity." *BioScience* 47(11): 747-757.

Pimentel, D., C. Harvey, P. Resosudarmo, K. Sinclair, D. Kurz, M. McNair, et al. 1995. "Environmental and Economic Costs of Soil Erosion and Conservation Benefits." *Science* 267: 1117–1123.

Pinzon, J., M. E. Brown, and C. J. Tucker. 2005. "Satellite time series correction of orbital drift artifacts using empirical mode decomposition." In: N. Huang (Editor), Hilbert-Huang Transform: Introduction and Applications, pp. 167-186.

Pitman, M. and A. Läuchli. 2004. Global Impact of Salinity and Agricultural Ecosystems. In: Läuchli, A. and U. Lüttge (eds.): Salinity: Environment - Plants – Molecules. Pp.3-20.

Place, F., and T. Hazell. 1993. "Productivity Effects on Indigenous Land Tenure Systems in Sub-Saharan Africa." *American Journal of Agricultural Economics* 75 (1): 10–19.

Place, F., and K. Otsuka. 2002. "Land Tenure Systems and Their Impacts on Agricultural Investments and Productivity in Uganda." *Journal of Development Studies* 38(6): 105–128.

Platteau, J. P. 1996. "The Evolutionary Theory of Land Rights as Applied to Sub-Saharan Africa. A Critical Assessment." *Development and Change* 27 (1): 29–86.

Posthumus, H., and J. de Graaf. 2005. "Cost-benefit Analysis of Bench Terraces: A Case Study in Peru." *Land Degradation and Development* 16: 1–11.

Pretty, J. B., C. Brett, D. Gee, R.E. Hine, C.F. Mason,J.I.L. Morison, H. Raven, M.D. Rayment, and G. van der Bijl 2000. "An Assessment of the Total External Costs of UK agriculture." *Agricultural Systems* 65 (2): 113–136.

Prince, S. 1991. "A Model of Regional Primary Production for Use with Coarse-Resolution Satellite Data." *International Journal of Remote Sensing* 12: 1313–1330.

Prince, S., I. Becker-Reshef, and K. Rishmawi. 2009. "Detection and Mapping of Long-Term Land Degradation Using Local Net Production Scaling: Application to Zimbabwe." *Remote Sensing of Environment* 113: 1046–1057.

Prince, S., K. Wessels, C. Tucker, and S. Nicholson. 2007. "Desertification in the Sahel: a reinterpretation of a reinterpretation." *Global Change Biology* 13: 1308–1313.

Qamar, M. K. 2005. *Modernizing National Agricultural Extensions System: A Practical Guide for Policy-Makers of Developing Countries.* Rome: Food and Agriculture Organization of the United Nations.

Randall, A. 1999. "Why Benefits and Costs Matter." *Choices* Second Quarter 1999: 38-42.

Ranjbar G.H., S.A.M. Cheraghi, A. Anagholi, G. L. Ayene, M.H. Rahimian, and M. Qadir. 2008. Introduction of salt-tolerant wheat, barley, and sorghum varieties in saline areas of the Karkheh River Basin. Proceedings of the CGIAR challenge program on water and food: 2nd International Forum (VII) CGIAR challenge program on water and food: 2nd International Forum. 10-14 Nov 2008, Addis Ababa, Ethiopia. p. 190-193.

Reardon, T., and S. Vosti. 1995. "Links between Rural Poverty and the Environment in Developing Countries: Asset Categories and Investment Poverty." *World Development* 23 (9): 1495–1506.

Reij, C., G. Tappan, and M. Smale. 2008. "Re-Greening the Sahel: Farmer-Led Innovation in Burkina Faso and Niger." In *Millions Fed. Proven Successes in Agricultural Development*, edited by D. Spielman and R. Pandya-Lorch, 53–58. Washington, DC: International Food Policy Research Institute.

Renard, K. G., G. R. Foster, G. A. Weesies, and J. P. Porter. 1991." RUSLE: Revised Universal Soil Loss Equation." *Journal of Soil and Water Conservation* 46: 30–33.

Requier-Desjardins, M. and M. Bied-Charreton 2006. *Investing in the Recovery of Arid Lands.* Working Document Presented at the International Workshop on the Cost of Inaction and Investment Opportunities in Dry, Arid, Semiarid and Subhumid Areas, Comité Scientifique Français de la Désertification (CDFD), Rome, December 4–5.

Requier-Desjardins, M. 2006. "The Economic Costs of Desertification: A First Survey of Some Cases in Africa." *International Journal of Sustainable Development* 9 (2): 199–209.

Requier-Desjardins, M., B. Adhikari, and S. Sperlich. 2011. "Some Notes on the Economic Assessment of Land Degradation." *Land Degradation and Development* 22: 285-298.

Reynolds, J., M. Stafford Smith, E. Lambin, and B. E. Turner. 2007. "Global Desertification: Building a Science for Drylands Development." *Science* 316: 847–851.

Ribaudo, M. O. 1986. "Consideration of Offsite Impacts in Targeting Soil Conservation Programs." *Land Economics* 62 (4): 402–411.

Richards, M. 1997. "The Potential for Economic Valuation of Watershed Protection in Mountainous Areas: A Case Study from Bolivia." *Mountain Research and Development* 17 (1): 19–30.

Robbins, L. H. 1984. "Late Prehistoric Aquatic and Pastoral Adaptations West of Lake Turkana, Kenya." In *From Hunters to Farmers: The Causes and Consequences of Food Production in Africa,* edited by J. D. Clark and S. Brandt, S206–211. Berkeley: University of California Press.

Rockström, J., J. Barron, and P. Fox. 2003. "Water Productivity in Rain-Fed Agriculture: Challenges and Opportunities for Smallholder Farmers in Drought-Prone Tropical Agroecosystems." In *Water Productivity in Agriculture: Limits and Opportunities for Improvement:* 145-162.

Rockström, J., W. Steffen, K. Noone, Å. Persson, F. Chapin, E. Lambin, et al. 2009. "A Safe Operating Space for Humanity." *Nature* 461: 472–475.

Safriel, U. N. 2007. "The Assessment of Global Trends in Land Degradation." In *Climate and Land Degradation,* edited by M. V. Sivakumar and N. Ndiang´ui, 2–38. Springer-Verlag, Berlin, Heidelberg, Germany

Safriel, U. and Z. Adeel. 2005. "Development Paths of Drylands: Thresholds and Sustainability." *Sustainability Science* 3: 117-123.

Safriel, U. N., and Z. Adeel 2005. "Dryland Systems." In *Ecosystems and Human Well-being: Current State and Trends.* Vol. 1, edited by R. Hassan, R. Scholes, & N. Ash, 623–662. Washington, DC: Island Press.

Sanwal, M. 2004. "Trends in Global Environmental Governance: The Emergence of a Mutual Supportiveness Approach to Achieve Sustainable Development." *Global Environmental Politics* 4 (4): 16–22.

Sauer, J., and H. Tchale. 2006. Alternative soil fertility management options in Malawi—an economic analysis. Paper presented at the International Association of Agricultural Economists Conference, Gold Coast, Australia, August 12–18.

Scherr, S., and P. Hazell. 1994. *Sustainable Agricultural Development Strategies in Fragile Lands.* Environment and Production Technology Division Discussion Paper, no. 1. Washington, DC: International Food Policy Research Institute.

Scherr, S.J., and S. Yadav. 1996. Land Degradation in the Developing World: Implications for Food, Agriculture, and the Environment to 2020. Washington, DC: International Food Policy Research Institute.

Scherr, S. J. 1999. Soil Degradation: A Threat to Developing-Country Food Security by 2020, food, Agriculture and the Environment. Washington, DC: International Food Policy Research Institute.

———. 2000. "Downward Spiral? Research Evidence on the Relationship between Poverty and Natural Resource Degradation." *Food Policy* 25 (4): 479–498.

Schlager, E., and E. Ostrom. 1992. "Property-Rights Regimes and Natural Resources." *Land Economics* 68 (3): 249–262.

Schneider, A., M. Friedl, and D. Potere. 2010. "Mapping Global Urban Areas using MODIS 500-m Data: New Methods and Datasets Based on 'Urban Ecoregions.'" *Remote Sensing of Environment* 114: 1733–1746.

Schuyt, K. D. 2005. "Economic Consequences of Wetland Degradation for Local Populations in Africa." *Ecological Economics* 53: 177–190.

Schwilch, G., B. Bestelmeyer, S. Bunning, W. Critchley, K. Kellner, H. P. Liniger, et al. (2009). "Lessons from Experiences in Monitoring and Assessment of Sustainable Land Management (MASLM)." Paper presented at the First Scientific UNCCD Conference on Monitoring and Assessment of Sustainable Land Management. Buenos Aires, Argentina.

Schwilch, G., H. Liniger, and G. van Lynden. 2004. "Towards a Global Map of Soil and Water Conservation Achievements: A WOCAT Initiative." In *Conserving Soil and Water for Society: Sharing Solutions* ISCO 2004 -

13th International Soil Conservation Organisation Conference , Brisbane, Australia, July 2004

SEDAC (Socioeconomic Data and Application Center). 2010. "National Boundaries and Gridded Population of the World, 2005 Population Density." http://sedac.ciesin.columbia.edu/gpw/abgerufen.

Shiferaw, B., and T. Holden. 1998. "Resource Degradation and Adoption of Land Conservation Technologies in the Highlands of Ethiopia: A Case Study of Andit Tid, North Sheawa." *Agricultural Economics* 18: 233–247.

———. 1999. "Soil Erosion and Smallholders' Conservation Decisions in the Highlands of Ethiopia." *World Development* 27 (4): 739–752.

———. 2000. "Policy Instruments for Sustainable Land Management: The Case of Highland Smallholders in Ethiopia." *Agricultural Economics* 22: 217–232.

———. 2001. "Farm-Level Benefits to Investments for Mitigating Land Degradation: Empirical Evidence from Ethiopia." *Environment and Development Economics* 6: 335–358.

———. 2005. "Assessing the Economic and Environmental Impacts of Conservation Technologies." In *Natural Resource Management in Agriculture. Methods for Assessing Economic and Environmental Impacts,* edited by B. Shiferaw, H. A. Freeman, and S. M. Swinton, S269–294. Oxfordshire, UK: CABI Publishing.

Shiferaw, B., S. T. Holden, and J. Aune. 2000. "Population pressure, Poverty and Incentives for soil conservation in Ethiopia: a bio-economic modeling approach." Paper presented at the Second International Conference on Environment and Development, Stockholm, 6–8 September, 2000.

Shiferaw, B., S. T. Holden, and J. Aune. 2001. Population pressure and land degradation in the Ethiopian highlands: a bio-economic model with endogenous soil degradation. In *Economic Policy Reforms and Sustainable Land Use – Recent Advances in Quantitative Analysis for Developing Countries,* edited by N. Heerink, H. van Keulen, and M. Kuiper. Springer Verlag, Berlin.

Shively, G. E. 1999. "Risks and Returns from Soil Conservation: Evidence from Low-Income Farms in the Philippines." *Agricultural Economics* 21: 53–67.

Smale, M., and T. Jayne. 2008. "Breeding an 'Amaizing' Crop Improved Maize in Kenya, Malawi, Zambia, and Zimbabwe." In *Millions Fed. Proven Successes in Agricultural Development*, edited by D. Spielman and R.

Pandya-Lorch. Washington, DC: International Food Policy Research Institute.

Smaling, E.M.A., L.O. Fresco, and A. de Jager. 1996. "Classifying, monitoring and improving soil nutrient stocks and flows in African agriculture." *Ambio* 25: 492-496.

Smedema L. K., S. Abdel-Dayem and W.J. Ochs. 2000. Drainage and Agricultural Development *Irrigation and Drainage Systems* 14(3):223-235.

Solorzano, R., R. De Camino, R. Woodward, J. Tosi, V. Watson, A. Vasquez, C. Villalobos, and J. Jime´nez. 1991. "Accounts Overdue: Natural Resource Depreciation in Costa Rica." World Resources Institute, Washington, DC.

Sonneveld, B. G. 2002. "Land Under Pressure: The Impact of Water Erosion on Food Production in Ethiopia." PhD thesis, Amsterdam University.

———. 2003. "Formalizing Expert Judgments in Land Degradation Assessment: A Case Study for Ethiopia." *Land Degradation and Development* 14: S347–361.

Sonneveld, B. G., and D. L. Dent. 2009. "How Good Is GLASOD?" *Journal of Environmental Management* 90 (1): 274–283.

Spooner, B. 1987. "The Paradoxes of Desertification." *Desertification Control Bulletin* 15: 40–45.

Sterk G. 2003. Causes, consequences and control of wind Erosion in Sahellan Africa: A review. *Land Degradation & Development*14: 95-108.

Stern, N. 2006. Stern Review: The Economics of Climate Change. London: UK Treasury.

Stiglitz, J. E., A. Sen, and J.-P. Fitoussi. 2009. Report by the Commission on the Measurement of Economic Performance and Social Progress.

Stocking, M. 1986. *The Cost of Soil Erosion in Zimbabwe in Terms of the Loss of Three Major Nutrients.* Consultant's Working Paper No. 3, Soil Conservation Programme. Rome: FAO Land and Water Division.

Stocking, M., and M. Murnaghan. 2005. Land Degradation: Guidelines for Field Assessment. London, UK.

Stockle, C.O., M. Donatelli and R. Nelson. 2003. CropSyst, a cropping systems simulation model. *European Journal of Agronomy* 18 (3-4): 289-307.

Stoorvogel, J.J., E. M. A. Smaling und B. H. Janssen. "Calculating soil nutrient balances in Africa at different scales." *Nutrient Cycling in Agroecosystems* 35(3): 227-235.

Stoorvogel, J. J., and E. M. Smaling. 1990. Assessment of Soil Nutrient Depletion in Sub-Saharan Africa: 1983–2000. Report 28. Wageningen, The Netherlands.

Stulina G., M.R. Cameira and L.S. Pereira. 2005. Using RZWQM to search improved practices for irrigated maize in Fergana, Uzbekistan. *Agricultural Water Management* 77(1-3):263-281.Sutcliffe, J.P. 1993. "Economic Assessment of Land Degradation in the Ethiopia Highlands. Addis Ababa." Natural Conservation Strategy Secretariat Ministry of Planning and Economic Development.

Syers, J. K., J. Lingard, C. E. Pieri, and G. Faure. 1993. Sustainable Land Management for the Semiarid and Subhumid tropics. *Ambio* 25 (8): 484–491.

Symeonakis, E., and N. Drake. 2004. "Monitoring Desertification and Land Degradation over Sub-Saharan Africa." *International Journal of Remote Sensing* 25 (3): 573–592.

Tan, Z. X., R. Lal, and K. D. Wiebe. 2005. "Global Soil Nutrient Depletion and Yield Reduction." *Journal of Sustainable Agriculture* 26 (1): 123–146.

Taylor, D.B. and D. L. Young. 1985. "The Influence of Technological Progress on the Long-Run Farm-Level Economics of Soil Conservation." Western Journal for Agricultural Economics 10: 63-76.

TEEB (The Economics of Ecosystems and Biodiversity). 2010. Mainstreaming the Economics of Nature: A Synthesis of the Approach, Conclusions and Recommendations of TEEB. Malta.

Tegtmeier, E. M., and M.D. Duffy. 2004. "External Costs of Agricultural Production in the United States." *International Journal of Agricultural Sustainability* 2 (1): 1-20.

Tengberg, A., and S.-I. B. Torheim. 2007. "The Role of Land Degradation in the Agriculture and Environment Nexus." In *Climate and Land Degradation,* edited by M. Sivakumar and N. Ndiang´ui, 267–283). Heidelberg: Springer Verlag.

Tenge, A.J., J. De Graaff, and J. P. Hella. 2004. "Social and Economic Factors Affecting the Adoption of Soil And Water Conservation in West Usambara Highlands, Tanzania." *Land Degradation & Development* 15(2): 99–114.

TERRASTAT 2010. Land Resource Potential and Constraints Statistics at Country and Regional Level. Online at: Land resource potential and constraints statistics at country and regional level.

TerrAfrica. 2006. Assessment of the Nature and Extent of Barriers and Bottlenecks to Scaling Sustainable Land Management Investments throughout Sub-Saharan Africa. Tanzania: Strategic Investment Program. Unpublished TerrAfrica report.

Thomas, D., and N. Middleton. 1994. *Desertification: Exploding the Myth.* West Sussex: British Library.

Thornton, P. K., R. L. Kruska, N. Henninger, P. M. Kristjanson, R. S. Reid, F. Atieno, A. N. Odero, and T. Ndegwa. 2002. Mapping poverty and livestock in the developing world. ILRI (International Livestock Research Institute); Nairobi, Kenya.

Tibke G. 1988. Basic Principles of Wind Erosion Control Agriculture. *Ecosystems & Environment* 22-23: 103-122

Tiffen, M., M. Mortimore, and F. Gichuki. 1994. *More People, Less Erosion: Environmental Recovery in Kenya.* London: Wiley and Sons.

Toderich K.N., E.V. Shuyskaya, S. Ismail, L. Gismatullina, T. Radjabov, B.B. Bekhchanov, and D. Aralova. 2009. Phytogenic Resources of Halophytes of Central Asia and their Role for Rehabilitation of Sandy Desert Degraded Rangelands. *Journal of Land Degradation and Development*, 20 (4): 386-396.

Toulmin, C., and J. Quan. 2000. "Registering Customary Rights." In *Evolving Land Rights, Policy, and Tenure in Africa,* edited by J. Toulmin and J. Quan, 207–228. London: International Institute for Environment and Development, Natural Resource Institute.

Toulmin C. 2009. Securing Land And Property Rights in Sub-Saharan Africa: The role of local Institutions *Land Use Policy* 26 (1) 10–19.

Troeh, F. R., J. A. Hobbs and R. L. Donahue. 1991. Tillage practices for conservation. In: *Soil and Water Conservation* edited by F. R. Troeh, J. A. Hobbs, and R. L. Donahue, 2nd ed. Englewood Cliffs, NJ: Prentice-Hall.

Tucker, C.J., J.E. Pinzon, and M.E. Brown. 2004. Global Inventory Modeling and Mapping Studies. NA94apr15b.n11-VIg, 2.0, Global Land Cover Facility, University of Maryland, College Park, Maryland, 04/15/1994.

Tucker, C.J., J. E. Pinzon, M. E. Brown, D. Slayback, E. W. Pak, R. Mahoney, E. Vermote and N. El Saleous. 2005. An Extended AVHRR 8-km NDVI Data Set Compatible with MODIS and SPOT Vegetation NDVI Data. *International Journal of Remote Sensing* 26 (20) 4485-5598.

Turner, R. K., D. Pearce, and I. Bateman. 1994. *Environmental Economics.* London: Harvester Wheatsheaf.

Twomlow S., J. Rusike, and S. Snapp. 2001. Biophysical or Economic Performance – Which Reflects Farmer Choiceof Legume 'Best-Bets' in Malawi? Proceedings of the Seventh Eastern and Southern Africa Regional maize conference. 11-15 February:480-486.

Qadir M., S. Schubert, A. D. Noble, M. Saqib and Saifullah. 2006. Amelioration strategies for Salinity-Induced Land Degradation. CAB Reviews: Perspectives in Agriculture, Veterinary Science, Nutrition and Natural Resources 1(069):1-12.

UNCCD. 1996. United Nations Convention to Combat Desertification in Countries Experiencing Serious Drought and/or Desertification, Particularly in Africa. http://www.unccd.int/convention/text/convention.php

UN Data. 2009. "Online Energy Data." http://data.un.org/Data.aspx?d=EDATA&f=cmID%3aES.

UNEP (United Nations Environment Program). 1977. *Draft Plan of Action to Combat Desertification.* Nairobi.

———. 1992. *World Atlas of Desertification* . Sevenoaks, UK: Edward Arnold.

———. 1997a. "Putting Desertification on the Map: UNEP Launches World Atlas of Desertification." www.grida.no/news/press/1751.aspx.

———. 1997b. Word Atlas of Desertification. 2nd ed. London.

———. 2011. Green Economy. Pathways to Sustainable Development and Poverty Eradication. A Synthesis for Policy Makers. New York.

USDA-NRCS (U. S. Department of Agriculture, Natural Resources Conservation Science). 1998. "Global Desertification Vulnerability Map." http://soils.usda.gov/use/worldsoils/mapindex/desert.html

———. 1999. "Risk of Human-Induced Desertification Map." http://soils.usda.gov/use/worldsoils/mapindex/dsrtrisk.html.

Vagen, T. G. 2005. "Soil Carbon Sequestration in Sub-Saharan Africa: A Review." *Land Degradation and Development* 16: 53–71.

Vanclay, F. 2004. "Social principles for agricultural extension to assist in the promotion of natural resource management." *Australian Journal of Experimental Agriculture* 44 (3): 213 – 222.

Van de Graaff, R., and R. A. Patterson. 2001. "Explaining the Mysteries of Salinity, Sodicity, SAR, and ESP in On-Site Practice." Paper presented at the On-Site '01 Conference: Advancing On-site Wastewater Systems, University of Armidale, New England, September 25–27.

Van der Pol, F. 1992. "Soil Mining, an Unseen Contributor to Farm Income in Southern Mali." *Bulletin* 325. Royal Tropical Institute, Amsterdam.

Van Donk S. and E.L. Skidmore. 2003.Measurement and simulation of wind erosion, roughness degradation and residue decomposition on an agricultural field. *Surface Processes and Landforms* 28(11):1243–1258.

Vanlauwe B., J. Kihara, P. Chivenge, P. Pypers, R. Coe and J. Six. 2011. Agronomic use efficiency of N fertilizer in maize-based systems in sub-Saharan Africa within the context of integrated soil fertility management *Plant and Soil* 339(1-2):35-50.Van Lynden, G., and T. Kuhlmann. 2002. "Review of Degradation Assessment Methods." LADA e-mail conference, October 4–9.

Vanlauwe B. 2007. Integrated soil fertility management research at TSBF: The framework, the principles, and their application. Tropical Soil Biology and Fertility (TSBF). Nairobi Kenya.

Van Lynden, G., H. Liniger, and G. Schwilch. 2002. "The WOCAT Map Methodology: A Standardized Tool for Mapping Degradation and Conservation." 12th International Soil Conservation Organization Conference. Beijing: World Overview of Conservation Approaches and Technologies; Wageningen, The Netherlands: International Soil Reference and Information Centre; Bern, Switzerland: Centre for Development and Environment, Institute of Geography.

Venkataratnam, L., and T. Sankar. 1996. "Remote Sensing and GIS for Assessment, Monitoring, and Management of Degraded Lands." In *Surveillance des Sols dans l'Environment par Télédétection et Systèmes d'Information Géographiques (Monitoring Soils in the Environment with Remote Sensing and GIS),* edited by R. Escadafal, M. Mulders, and L. Thiombiano, 503–516. Ouagadougou, Burkina Faso: Office de la Recherche Scientifique et Technique d'Outre-Mer.

Vieth, G. R., H. Gunatilake, and L. J. Cox. 2001. "Economics of Soil Conservation. The Upper Mahaweli Watershed of Sri Lanka." *Journal of Agricultural Economics* 52 (1): 139–152.

Vlek, P. L., Q. B. Le, and L. Tamene. 2008. *Land Decline in Land-Rich Africa: A Creeping Disaster in the Making.* Rome: Consultative Group on International Agricultural Research, Science Council Secretariat.

———. 2010. "Assessment of Land Degradation, Its Possible Causes, and Threat to Food Security in Subsaharan Africa." In *Food Security and Soil Quality,* edited by R. Lal and B. A. Stewart, 57–86. Boca Raton, FL: CRC Press.

Vlek P.L., G. Rodriguez-Kuhl and R. Sommer. 2004. Energy use and CO_2 Production in Tropical Agriculture and Means and Strategies for Reduction or Mitigation. *Environment, Development and Sustainability* 6: 213–233.

Von Braun, J., H. de Haen, and J. Blanken. 1991. *Commercialization of Agriculture under Population Pressure: Effects on Production, Consumption, and Nutrition in Rwanda*. IFPRI Research Report 85. Washington, DC: International Food Policy Research Institute.

Voortman, R. L., B. G. Sonneveld, and M. A. Keyzer. 2000. African land ecology: Opportunities and constraints for agricultural development. Center for International Development Working Paper 37. Cambridge, Mass., U.S.A.: Harvard University

Vosti, S. A., and T. Reardon. 1997. Sustainability, Growth, and Poverty Alleviation. A Policy and Agroecological Perspective. Baltimore, MD: Johns Hopkins University Press.

Walker, T., and J. Ryan. 1990. *Village and Household Economies in India's Semi-Arid Tropics.* Baltimore: Johns Hopkins University Press.

Wallace, K. J. 2007. "Classification of Ecosystem Services: Problems and Solutions." *Biological Conservation* 139: 235–246.

Walker, D. J. 1982. "A Damage Function to Evaluate Erosion Control Economics." *American Journal of Agricultural Economics* 64:690-98.

Walsh, K.N. and M. Rowe. 2001, Biodiversity Increases Ecosystems' Ability to Absorb CO_2 and Nitrogen, <http://www.bnl.gov/bnlweb/pubaf/pr/2001bnlpr041101.htm.> (8/10/2002).

Walton P., R. Martinez, and A. Bailey. 1981. "A Comparison of Continuous and Rotational Grazing." *Journal of Range Management* 34 (1): 19–21.

Wane A. 2005. "Marchés de bétail du Ferlo (Sahel sénégalais) et comportements des ménages pastoraux." Colloque Société francaise d´économie rurale (SFER), Montpellier, France, November 7–9.

Warren, A. 2002. "Land Degradation Is Contextual." *Land Degradation and Development* 13: 449–459.

Way, S. A. 2006. "Examining the Links between Poverty and Land Degradation: From Blaming the Poor toward Recognizing the Rights of the Poor." In *Governing Global Desertification: Linking Environmental Degradation, Poverty, and Participation,* edited by P. Johnson, K. Mayrand, & M. Paquin, 27–41. Burlington, VT: Ashgate.

WDPA (World Database on Protected Areas). 2008. www.wdpa.org/.

Wessels, K. 2009. “Comments on ‘Proxy Global Assessments of Land Degradation’ by Bai et al. 2008.” *Soil Use and Management:* 91–92.

Wessels, K., S. Prince, J. Malherbe, J. Small, P. Frost, and D. Van Zyl. 2007. “Can Human-Induced Land Degradation Be Distinguished from the Effects of Rainfall Variability? A Case Study in South Africa.” *Journal of Arid Environments* 68: 271–297.

Whitaker, B. E. 1999. *Changing Opportunities: Refugees and Host Communities in Western Tanzania.* Journal of Humanitarian Assistance Working Paper 11. Geneva: United Nations High Commissioner for Refugees.

White, W. R. 2010. “World Water: Resources, Usage, and the Role of Man-Made Reservoirs.” Foundation for Water Resources. www.fwr.org. 60pp.

Wiggins, S., and O. Palma. 1980. *Acelhuate River Catchment Management Project, El Salvador: Cost Benefit Analysis of Soil Conservation.* Land Resources Development Centre, Project Report, no. 104. Surbiton, UK: Overseas Development Administration.

Williams, J. R., K. G. Renard, and P. T. Dyke. 1983. “EPIC: A New Method for Assessing Erosion’s Effect on Soil Productivity.” *Journal of Soil and Water Conservation* 38 (5): 381–386.

Williamson, O. 2000. “The New Institutional Economics: Taking Stock and Looking Ahead.” *Journal of Economic Literature* 38: 595–613.

Wint, W. and T. Robinson. 2007: Gridded Livestock of the World. FAO, Rome 2007.

Winters, P., C. Crissman, and P. Espinosa. 2004. “Inducing the adoption of conservation technologies: lessons from the Ecuadorian Andes.” *Environmental and Development Economics* 9: 695–719.

Wischmeier, W.H. 1976. “Use and Misuse of the universal soil loss equation.” *Journal of Soil and Water Conservation.* Vol. 13 (1): 5-9.

Wischmeier, W., and D. Smith. 1978. *Predicting Rainfall Erosion Losses: A Guide to Conservation planning* (Handbook 537). Washington, DC: U.S. Department of Agriculture.

WMO (World Meteorological Organization). 2005. *Climate and Land Degradation.* WMO Report No. 989. Geneva, Switzerland.

Woelcke, J. 2003. Bio-Economics of Sustainable Land Management in Uganda. Development Economics and Policy 37. Peter Lang Verlag. Frankfurt am Main, Germany.

World Bank. 2007. World Development Report: Agriculture for Development. Washington, DC.

———. 2008. *Uganda Sustainable Land Management Public Expenditure Review (SLM PER). Agriculture and Rural Development Unit (AFTAR).* Sustainable Development Department, Report45781-UG. Washington, DC.

———. 2009. *Impacts of Sustainable Land Management Programs on Land Management and Poverty in Niger.* World Bank Country Report 48230-NE. Washington DC.

———. 2010. Sub-Saharan Africa—Managing Land in a Changing Climate: An Operational Perspective for Sub-Saharan Africa. Sustainable Development Department, Africa Region, Report 54134-AFR. Washington, DC.

———. 2011. "Mali: Key Institutional, Financing, and Economic Elements for Scaling Up Sustainable Land and Water Management." Mimeo. World Bank, Washington, DC.

Yatich, T., A. Kalinganire, K. Alinon, J. C. Weber, J. M. Dakouo, O. Samaké, and S. Sangaré. 2008. *Moving Beyond Forestry Laws in Sahelian Countries.* World Agroforestry Center Policy Brief. Nairobi, Kenya.

Yesuf, M., A. Mekonnen, M. Kassie, and J. Pender. 2005. "Cost of Land Degradation in Ethiopia: A Critical Review of Past Studies." *Report, Environment for Development (EfD) World Bank. Gothenburg, Sweden.*

You, L. and S. Wood. 2006. An entropy approach to spatial disaggregation of agricultural production. *Agricultural Systems* 90: 329-347.

Young, A. 1993. "Land Degradation in South Asia: Its Severity, Causes, and Effects upon the People". Final report prepared for submission to the United Nations Economic and Social Council of the United Nations. Food and Agriculture Organization of the United Nations, United Nations Development Programme and United Nations Environment Programme, Rome.

Development Economics and Policy

Series edited by Franz Heidhues, Joachim von Braun and Manfred Zeller

Band 1 Andrea Fadani: Agricultural Price Policy and Export and Food Production in Cameroon. A Farming Systems Analysis of Pricing Policies. The Case of Coffee-Based Farming Systems. 1999.

Band 2 Heike Michelsen: Auswirkungen der Währungsunion auf den Strukturanpassungsprozeß der Länder der afrikanischen Franc-Zone. 1995.

Band 3 Stephan Bea: Direktinvestitionen in Entwicklungsländern. Auswirkungen von Stabilisierungsmaßnahmen und Strukturreformen in Mexiko. 1995.

Band 4 Franz Heidhues / François Kamajou: Agricultural Policy Analysis – Proceedings of an International Seminar, held at the University of Dschang, Cameroon on May 26 and 27 1994, funded by the European Union under the Science and Technology Program (STD). 1996.

Band 5 Elke M. Förster: Protection or Liberalization? A Policy Analysis of the Korean Beef Sector. 1996.

Band 6 Gertrud Schrieder: The Role of Rural Finance for Food Security of the Poor in Cameroon. 1996.

Band 7 Nestor R. Ahoyo Adjovi: Economie des Systèmes de Production intégrant la Culture de Riz au Sud du Bénin: Potentialités, Contraintes et Perspectives. 1996.

Band 8 Jenny Müller: Income Distribution in the Agricultural Sector of Thailand. Empirical Analysis and Policy Options. 1996.

Band 9 Michael Brüntrup: Agricultural Price Policy and its Impact on Production, Income, Employment and the Adoption of Innovations. A Farming Systems Based Analysis of Cotton Policy in Northern Benin. 1997.

Band 10 Justin Bomda: Déterminants de l'Epargne et du Crédit, et leurs Implications pour le Développement du Système Financier Rural au Cameroun. 1998.

Band 11 John M. Msuya: Nutrition Improvement Projects in Tanzania: Implementation, Determinants of Performance, and Policy Implications. 1998.

Band 12 Andreas Neef: Auswirkungen von Bodenrechtswandel auf Ressourcennutzung und wirtschaftliches Verhalten von Kleinbauern in Niger und Benin. 1999.

Band 13 Susanna Wolf (ed.): The Future of EU-ACP Relations. 1999.

Band 14 Franz Heidhues / Gertrud Schrieder (eds.): Romania – Rural Finance in Transition Economies. 2000.

Band 15 Katinka Weinberger: Women's Participation. An Economic Analysis in Rural Chad and Pakistan. 2000.

Band 16 Christof Batzlen: Migration and Economic Development. Remittances and Investments in South Asia: A Case Study of Pakistan. 2000.

Band 17 Matin Qaim: Potential Impacts of Crop Biotechnology in Developing Countries. 2000.

Band 18 Jean Senahoun: Programmes d'ajustement structurel, sécurité alimentaire et durabilité agricole. Une approche d'analyse intégrée, appliquée au Bénin. 2001.

Band 19 Torsten Feldbrügge: Economics of Emergency Relief Management in Developing Countries. With Case Studies on Food Relief in Angola and Mozambique. 2001.

Band 20 Claudia Ringler: Optimal Allocation and Use of Water Resources in the Mekong River Basin: Multi-Country and Intersectoral Analyses. 2001.

Band 21 Arnim Kuhn: Handelskosten und regionale (Des-)Integration. Russlands Agrarmärkte in der Transformation. 2001.

Band 22 Ortrun Anne Gronski: Stock Markets and Economic Growth. Evidence from South Africa. 2001.

Band 23 Patrick Webb / Katinka Weinberger (eds.): Women Farmers. Enhancing Rights, Recognition and Productivity. 2001.

Band 24 Mingzhi Sheng: Lebensmittelkonsum und -konsumtrends in China. Eine empirische Analyse auf der Basis ökonometrischer Nachfragemodelle. 2002.

Band 25 Maria Iskandarani: Economics of Household Water Security in Jordan. 2002.

Band 26 Romeo Bertolini: Telecommunication Services in Sub-Saharan Africa. An Analysis of Access and Use in the Southern Volta Region in Ghana. 2002.

Band 27 Dietrich Müller-Falcke: Use and Impact of Information and Communication Technologies in Developing Countries' Small Businesses. Evidence from Indian Small Scale Industry. 2002.

Band 28 Wolfram Erhardt: Financial Markets for Small Enterprises in Urban and Rural Northern Thailand. Empirical Analysis on the Demand for and Supply of Financial Services, with Particular Emphasis on the Determinants of Credit Access and Borrower Transaction Costs. 2002.

Band 29 Wensheng Wang: The Impact of Information and Communication Technologies on Farm Households in China. 2002.

Band 30 Shyamal K. Chowdhury: Institutional and Welfare Aspects of the Provision and Use of Information and Communication Technologies in the Rural Areas of Bangladesh and Peru. 2002.

Band 31 Annette Luibrand: Transition in Vietnam. Impact of the Rural Reform Process on an Ethnic Minority. 2002.

Band 32 Felix Ankomah Asante: Economic Analysis of Decentralisation in Rural Ghana. 2003.

Band 33 Chodechai Suwanaporn: Determinants of Bank Lending in Thailand: An Empirical Examination for the Years 1992 to 1996. 2003.

Band 34 Abay Asfaw: Costs of Illness, Demand for Medical Care, and the Prospect of Community Health Insurance Schemes in the Rural Areas of Ethiopia. 2003.

Band 35 Gi-Soon Song: The Impact of Information and Communication Technologies (ICTs) on Rural Households. A Holistic Approach Applied to the Case of Lao People's Democratic Re- public. 2003.

Band 36 Daniela Lohlein: An Economic Analysis of Public Good Provision in Rural Russia. The Case of Education and Health Care. 2003.

Band 37 Johannes Woelcke. Bio-Economics of Sustainable Land Management in Uganda. 2003.

Band 38 Susanne M. Ziemek: The Economics of Volunteer Labor Supply. An Application to Countries of a Different Development Level. 2003.

Band 39 Doris Wiesmann: An International Nutrition Index. Concept and Analyses of Food Insecurity and Undernutrition at Country Levels. 2004.

Band 40 Isaac Osei-Akoto: The Economics of Rural Health Insurance. The Effects of Formal and Informal Risk-Sharing Schemes in Ghana. 2004.

Band 41 Yuansheng Jiang: Health Insurance Demand and Health Risk Management in Rural China. 2004.

Band 42 Roukayatou Zimmermann: Biotechnology and Value-added Traits in Food Crops: Relevance for Developing Countries and Economic Analyses. 2004.

Band 43 F. Markus Kaiser: Incentives in Community-based Health Insurance Schemes. 2004.

Band 44 Thomas Herzfeld: *Corruption begets Corruption*. Zur Dynamik und Persistenz der Korruption. 2004.

Band 45 Edilegnaw Wale Zegeye: The Economics of On-Farm Conservation of Crop Diversity in Ethiopia: Incentives, Attribute Preferences and Opportunity Costs of Maintaining Local Varieties of Crops. 2004.

Band 46 Adama Konseiga: Regional Integration Beyond the Traditional Trade Benefits: Labor Mobility contribution. The Case of Burkina Faso and Côte d'Ivoire. 2005.

Band 47 Beyene Tadesse Ferenji: The Impact of Policy Reform and Institutional Transformation on Agricultural Performance. An Economic Study of Ethiopian Agriculture. 2005.

Band 48 Sabine Daude: Agricultural Trade Liberalization in the WTO and Its Poverty Implications. A Study of Rural Households in Northern Vietnam. 2005.

Band 49 Kadir Osman Gyasi: Determinants of Success of Collective Action on Local Commons. An Empirical Analysis of Community-Based Irrigation Management in Northern Ghana. 2005.

Band 50 Borbala E. Balint: Determinants of Commercial Orientation and Sustainability of Agricultural Production of the Individual Farms in Romania. 2006.

Band 51 Pamela Marinda: Effects of Gender Inequality in Resource Ownership and Access on Household Welfare and Food Security in Kenya. A Case Study of West Pokot District. 2006.

Band 52 Charles Palmer: The Outcomes and their Determinants from Community-Company Contracting over Forest Use in Post-Decentralization Indonesia. 2006.

Band 53 Hardwick Tchale: Agricultural Policy and Soil Fertility Management in the Maize-based Smallholder Farming System in Malawi. 2006.

Band 54 John Kedi Mduma: Rural Off-Farm Employment and its Effects on Adoption of Labor Intensive Soil Conserving Measures in Tanzania. 2006.

Band 55 Mareike Meyn: The Impact of EU Free Trade Agreements on Economic Development and Regional Integration in Southern Africa. The Example of EU-SACU Trade Relations. 2006.

Band 56 Clemens Breisinger: Modelling Infrastructure Investments, Growth and Poverty Impact. A Two-Region Computable General Equilibrium Perspective on Vietnam. 2006.

Band 57 Meike Wollni: Coping with the Coffee Crisis. An Analysis of the Production and Marketing Performance of Coffee Farmers in Costa Rica. 2007.

Band 58 Franklin Simtowe: Performance and Impact of Microfinance. Evidence from Joint Liability Lending Programs in Malawi. 2008.

Band 59 Xiangping Jia: Credit Rationing and Institutional Constraint. Evidence from Rural China. 2008.

Band 60 Holger Seebens: The Economics of Gender and the Household in Developing Countries. 2009.

Band 61 Stephan Piotrowski: Land Property Rights and Natural Resource Use. An Analysis of Household Behavior in Rural China. 2009.

Band 62 Sebastian M. Scholz: Rural Development through Carbon Finance. Forestry Projects under the Clean Development Mechanism of the Kyoto Protocol. Assessing Smallholder Participation by Structural Equation Modeling. 2009.

Band 63 Jakob Rupert Friederichsen: Opening Up Knowledge Production through Participatory Research? Agricultural Research for Vietnam's Northern Uplands. 2009.

Band 64 Olivier Ecker: Economics of Micronutrient Malnutrition. The Demand for Nutrients in Sub-Saharan Africa. 2009.

Band 65 Julia Johannsen: Operational Assessment of Monetary Poverty by Proxy Means Tests. 2009

Band 66 Ephraim Nkonya / Nicolas Gerber / Philipp Baumgartner / Joachim von Braun / Alessandro De Pinto / Valerie Graw / Edward Kato / Julia Kloos / Teresa Walter: The Economics of Land Degradation. Toward an Integrated Global Assessment. 2011.

www.peterlang.de

www.ingramcontent.com/pod-product-compliance
Ingram Content Group UK Ltd.
Pitfield, Milton Keynes, MK11 3LW, UK
UKHW021359190726
13851UKWH00015B/133